Intelligent Systems Reference Library

Volume 145

Series editors

Janusz Kacprzyk, Polish Academy of Sciences, Warsaw, Poland
e-mail: kacprzyk@ibspan.waw.pl

Lakhmi C. Jain, University of Canberra, Canberra, Australia;
Bournemouth University, UK;
KES International, UK
e-mail: jainlc2002@yahoo.co.uk; jainlakhmi@gmail.com
URL: http://www.kesinternational.org/organisation.php

The aim of this series is to publish a Reference Library, including novel advances and developments in all aspects of Intelligent Systems in an easily accessible and well structured form. The series includes reference works, handbooks, compendia, textbooks, well-structured monographs, dictionaries, and encyclopedias. It contains well integrated knowledge and current information in the field of Intelligent Systems. The series covers the theory, applications, and design methods of Intelligent Systems. Virtually all disciplines such as engineering, computer science, avionics, business, e-commerce, environment, healthcare, physics and life science are included. The list of topics spans all the areas of modern intelligent systems such as: Ambient intelligence, Computational intelligence, Social intelligence, Computational neuroscience, Artificial life, Virtual society, Cognitive systems, DNA and immunity-based systems, e-Learning and teaching, Human-centred computing and Machine ethics, Intelligent control, Intelligent data analysis, Knowledge-based paradigms, Knowledge management, Intelligent agents, Intelligent decision making, Intelligent network security, Interactive entertainment, Learning paradigms, Recommender systems, Robotics and Mechatronics including human-machine teaming, Self-organizing and adaptive systems, Soft computing including Neural systems, Fuzzy systems, Evolutionary computing and the Fusion of these paradigms, Perception and Vision, Web intelligence and Multimedia.

More information about this series at http://www.springer.com/series/8578

Halina Kwaśnicka · Lakhmi C. Jain
Editors

Bridging the Semantic Gap in Image and Video Analysis

 Springer

Editors
Halina Kwaśnicka
Department of Computational Intelligence
Wrocław University of Science
and Technology
Wrocław
Poland

Lakhmi C. Jain
Faculty of Education, Science,
Technology and Mathematics
University of Canberra
Canberra, ACT
Australia

ISSN 1868-4394 ISSN 1868-4408 (electronic)
Intelligent Systems Reference Library
ISBN 978-3-030-08879-8 ISBN 978-3-319-73891-8 (eBook)
https://doi.org/10.1007/978-3-319-73891-8

This Springer imprint is published by the registered company Springer International Publishing AG
part of Springer Nature
The registered company address is: Gewerbestrasse 11, 6330 Cham, Switzerland

Preface

Tremendous advances in intelligent techniques and wide applications of vision systems have resulted in an exponential increase in research in the field of image interpretation, also called as semantic analysis of images.

Semantic gap is a challenging area for research. The semantic gap is a difference between low-level features extracted from the image and the high-level semantic meanings that people can recognize on the image. The scene understanding is the highest level of image processing. The result strongly depends on the lower-level vision techniques, such as features selection, segmentation, objects recognition, and so on. The chapters in the book deal with different stages of image processing influenced on bridging the semantic gap.

Chapter 1 introduces the semantic gap in image and video analysis. Chapter 2 reports and compares a selected set of feature extractors. These extractors are chosen due to their use in a wide number of smart machine vision systems. Chapter 3 is on promising conformity of segmentation results with semantic image interpretation. Relations between semantics-based image annotation and SIMSER features are investigated. Chapter 4 is also on segmentation. It describes the active partitions technique which is a generalization of known active contour approach. Chapter 5 reports 3D object recognition techniques and object model. Chapter 6 demonstrates that structured video annotations can be efficiently queried manually or programmatically and can be used in scene interpretation, video understanding, and content-based video retrieval. Chapter 7 presents an overview of deep learning in semantic gap at different levels of image processing.

The book is directed to the researchers, practitioners, students, and professors in the field of semantic image processing, multimedia processing, and deep learning applied to semantic gap. The book was meant to indicate the direction of research for all who seek an answer to the question: how to overcome the semantic gap in images and video analysis.

We wish to express our gratitude to the authors and reviewers for their contributions. The assistance offered by the Springer-Verlag is acknowledged.

Wrocław, Poland   Halina Kwaśnicka
Canberra, Australia   Lakhmi C. Jain

Contents

Chapter 1
Semantic Gap in Image and Video Analysis: An Introduction

Halina Kwaśnicka and Lakhmi C. Jain

Abstract The chapter presents a brief introduction to the problem with the semantic gap in content-based image retrieval systems. It presents the complex process of image processing, leading from raw images, through subsequent stages to the semantic interpretation of the image. Next, the content of all chapters included in this book is shortly presented.

1.1 Introduction

The problem of the semantic gap is crucial and is seen in many tasks of image analysis, as Content-Based Image Retrieval (CBIR) or Automatic Image Annotation (AIA). The semantic gap is a lack of correspondence between the low-level information extracted from an image and the interpretation that the image has for a user. How to transform the features computed from raw image data to the high-level representation of semantics carried out by that image is still the open problem. This problem exists despite the observed intensive research with the use of different approaches to solving, or at least narrowing, the semantic gap in image analysis, especially in image retrieval. This gap is perceived as a barrier to image understanding. Some researchers claim that the understanding of how humans perceive images should be helpful [1, 2]. A typical CBIR method is a query-by-example system. In real life application finding an image as an appropriate users query is hard [3]. Easier and more intuitive is to describe the intended image by some keywords. Combining different media, like images, text,

H. Kwaśnicka (✉)
Department of Computational Intelligence, Wroclaw University of Science and Technology, Wroclaw, Poland
e-mail: halina.kwasnicka@pwr.edu.pl

L. C. Jain
Founder, KES International, Leeds, UK

L. C. Jain
Faculty of Science, Technology and Mathematics, University of Canberra, Canberra, Australia
e-mail: jainlakhmi@gmail.com; jainlc2002@yahoo.co.uk

© Springer International Publishing AG, part of Springer Nature 2018   1
H. Kwaśnicka and L. C. Jain (eds.), *Bridging the Semantic Gap in Image and Video Analysis*, Intelligent Systems Reference Library 145,
https://doi.org/10.1007/978-3-319-73891-8_1

video, sound, into one application is a subject of Multimedia Information Retrieval. It is also the widely developed field of research.

The output of CBIR systems is a ranked list of images; the images are ordered according to their similarity to the users query image. However, similarity is measured using low-level features extracted from images; this causes that returned images often do not meet users expectations, similarity based on low-level features do not correspond the human perception of similarity. Research on how human perception is working is intensively developed, one can expect that their results will be useful in bridging the semantic gap [4–9]. Authors of [9] try to model of human cortical function aiming simulation of the time-course of cortical processes of understanding meaningful, concrete words. The different parts of the cortex are responsible for general and selective, or category-specic, semantic processing. In [5] authors studied the humans and automatic perception of orientation of color photographic images. They concluded that the interaction with the human observers allows defining sky and people as the most important cues used by humans at various image resolutions. These and other results in the field of understanding human perception can be a hint for the creators of computer systems understanding images. Some researchers focus on developing a computer system that mimics the perceptual ability of people [10]. Such systems try to consider knowledge about the structure and the surrounding environment of a scene.

An analysis of the perception of images by man suggests that computer vision systems should also take into account some knowledge. The computer systems require acquired knowledge at different levels. To explain it let us see on vision systems from three perspectives: knowledge, algorithmic and implementation perspectives. From the implementation perspective, the used programming languages and computer hardware can be considered; this is not interesting for us here. The algorithmic perspective is essential—we have to decide the way of representing the relevant information, also the most suitable algorithms for use. The most interesting perspective is the knowledge perspective. Here, the questions could concern the knowledge that enters a process, the knowledge obtained in the process, constraints determining the process, and others.

An image (a scene) corresponds to basic properties of real-world. The next processing step uses physics, photometry, and so on. Further processing requires models of objects to be recognized, models of situations and common sense knowledge (see Fig. 1.1).

Information derived from primitive features, extracted from images, is the low-level knowledge. The semantic relationships and patterns, gathered by knowledge discovering methods, are the second level of knowledge [10]. Gathering such knowledge requires considering the correlation between the low-level information with the interpretation of concepts related to domain knowledge. Machine learning has to recognize complex structural relations between visual data and the semantic interpreted by human observing the considered scene.

In real-life use of CBIR systems, often a user can have a problem with finding a query image that matches the user's intent [11]. Finding the perfect image from a collection could be an example of such situation. It would then be much easier to

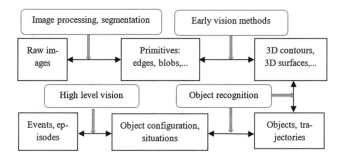

Fig. 1.1 From raw image to image understanding—a schema of processing

describe the desired image using text. The authors of [12], distinguish four scenarios depending on available information for creating CBIR: caption; annotation, tag, keyword; full MPEG-7 annotation. The potential scenarios are: only images; images with captions; images with captions and annotations, tags, keyword; and images with all mentioned descriptions. The authors propose different corresponding tasks for these scenarios such as rule induction for semantic class refinement, use Knowledge-Based System to infer object association or structural projection MPEG-7 representation and index building.

Multimodal CBIRs, i.e., taking into consideration visual, textual and audio features are growing in popularity. How to exploit the visual content of images in the CBIR systems is strongly developed, but there are other subjects worth the attention of researchers. Li et al. present a survey of researches on three problems connected with the semantic gap bridging: image tag assignment, refinement, and tag-based image retrieval [13]. The tag relevance to the visual content of an image hardly influences the quality of CBIR.

As it was mentioned earlier, the subject of semantic gap in the field of content-based image retrieval is intensively studied. The very interesting survey is presented in [14]. Authors comprehensively present achievements in particular steps of the CBIR systems, starting from the framework of CBIR, by image preprocessing, feature extraction, learning system, benchmark datasets, similarity matching, relevance feedback, up to the evaluation of performance and visualization. The authors also indicate some key issues that influence the CBIR. They pointed out as still open problems: representation of images with a focus on local descriptors; automatic image annotation; image indexing to reduce dimensionality; deep learning approach; description of ideal image datasets; re-ranking approaches as post-processing; visualization aspects.

An interesting approach is presented in [15]. The authors extend the latent semantic word and object models, to the latent semantic word, object and part models. The premise of this approach was the fact that not only similarity of semantic of words and semantic of images is important to the CBIR task. Also complex semantic relations within each modality, e.g., there are similar relations in the text to the relation between objects: *object A is a part of object B* and *object B is an instance of object*

C. They developed models able to learn these types of semantic relations across and within modalities simultaneously, using ImageNet and WordNet sources.

Variety of approaches have been developed to improve the CBIR systems that would be able to return the most relevant images with maximum user satisfaction [16–19]. Also, numerous papers containing a survey of the CBIR systems have been published, i.e., [13, 14, 20, 21]. In this book, some chapters present interesting approaches at the different level of CBIR systems and one chapter dedicated to applications of deep learning to bridge semantic gap. We have noticed a lack of survey dedicated to this new learning paradigm applied to image understanding, and the last chapter fills this gap.

1.2   Chapters Included in the Book

Chapter 2 presents a comparative study of the most used and popular low-level local feature extractors in a smart image and video analysis. An overview of different extractors is the first part of the chapter. The authors highlighted the main theoretical differences among the different extractors. A comprehensive study has been performed with use the Freiburg-Berkeley Motion Segmentation (FBMS-59) dataset. The robustness and behavior of compared extractors are discussed. The observations about the matching process are also outlined.

Chapter 3 is dedicated to image segmentation. The author claims that reliable segmentation algorithms, extracting as accurately as possible, regions with a certain level of semantic uniformity significantly improve the automatic annotation of an image. The developed segmentation technique is based on scale-insensitive maximally stable extremal regions (SIMSER) features a generalization of the popular MSER features, which is rather useless in semantic image segmentation. The chapter describes the experimental study of relations between semantics based image annotation and SIMSER features, focusing on color images.

Chapter 4 shows a generalization of known active contour technique, namely active partitions. The proposed approach can be applied to more sophisticated image content representations than raw pixel data. The reduction of search space enables to use evolutionary computations, less sensitive or invariant to the choice of initial solutions. The author demonstrates the flexibility of the proposed approach; it can be applied to both global and local image analysis.

Chapter 5 deals with 3D object recognition in RGB-D images, in indoor autonomous robotics. The proposed framework integrates solutions for: generic object representation; trainable transformations between abstraction levels; reasoning under uncertain and partial data; optimized model-to-data matching; efficient search strategies. As such, the framework is an application-independent generic model based. It was verified in robot vision scenarios. The approach allows to identify what kind of knowledge is needed and to utilize existing meta-level knowledge to learn concept types instead of memorizing individual instances. An interesting feature of the proposed framework is decomposition of an object into simpler elements, named parts. The authors confirmed experimentally that the approach might easily be adapted to multiple scenarios.

Chapter 6 concerns efficient automated mechanisms for processing video contents. The vast gap between what humans can comprehend based on cognition, knowledge, and experience, and what computer systems can obtain from signal processing, causes the subject very difficult. On the other hand, the increasing popularity and ubiquity of videos need efficient automated mechanisms for processing video contents. The spatiotemporal annotation of complex video scenes, in the form interpretable for machines, can be obtained by fusion of structured descriptions with textual and audio descriptors. This annotation can be used in scene interpretation, video understanding, and content-based video retrieval.

Chapter 7 focuses on how deep learning can be used in bridging the semantic gap in the content-based image retrieval. The chapter briefly presents the traditional approaches and introduces into deep learning, methods and deep models useful in CBIR. The authors distinguished three basic structure levels for scene interpretation using deep learning; they are feature level, common sense knowledge level, and inference level. The chapter presents the applications of deep learning at the particular levels of CBIR. Finally, the application deep models in bridging the semantic gap are summed in a table, and the growing popularity of DL in image analysis is shown.

1.3 Conclusion

The chapter provides some problems connected with a gap between automatic image interpretation and how human perceive the semantic content of an image. Steps of image processing from raw image to semantic image interpretation are presented. Each step influences the result of CBIR systems. From the semantic gap bridging point of view, the most interesting seems to be a knowledge level of image analysis. However, it strongly depends on the lower levels. A raw image reflects basic real-world properties. Features extracted from a raw image strongly influence the further process, and by this, the final results. Deep models are becoming increasingly popular and are rapidly developed. They deal with complicated tasks such as choosing the suitable set of features. Instead, they learn the feature. Deep models release a human from the need to define features and algorithms of image processing; they are worth developing.

References

1. Alzubaidi, M.A., Narrowing the semantic gap in natural images. In: 5th International Conference on Information and Communication Systems (ICICS), Irbid, 2014, pp. 1–6 (2014). https://doi.org/10.1109/IACS.2014.6841972

2. Alzubaidi, M.A.: A new strategy for bridging the semantic gap in image retrieval. Int. J. Comput. Sci. Eng. (IJCSE) **14**(1) (2017)
3. Jaimes, A., Christel, M., Gilles, S., Sarukkai, R., Ma, W.-Y.: Multimedia information retrieval: what is it, and why isn't anyone using it? In: Proceeding MIR 2005, Proceedings of the 7th ACM SIGMM International Workshop on Multimedia Information Retrieval, Hilton, Singapore, pp. 3–8 (2005)
4. Luke, K.-K, Liu, H.-L, Wai, Y.-Y., Wan, Y.-L., Tan, L.H.: Functional anatomy of syntactic and semantic processing in language comprehension. Hum. Brain Mapp. **16**(3), 133–145 (2002)
5. Luo, J., Crandall, D., Singhal, A., Boutell, M., Gray, R.T.: Psychophysical study of image orientation perception. Spat. Vis. **16**(5), 429457 (2003)
6. Friedrich R.M., Friederici A.D.: Mathematical logic in the human brain: semantics. PLoS ONE **8**(1), e53699 (2013). https://doi.org/10.1371/journal.pone.0053699
7. Rommers, J., Dijkstra, T., Bastiaansen, M.: Context-dependent semantic processing in the human brain: evidence from idiom comprehension. J. Cogn. Neurosci. **25**(5), 762–776 (2013)
8. Mitchell, D.J., Cusack, R.: Semantic and emotional content of imagined representations in human occipitotemporal cortex. Sci. Rep. **6**, 20232 (2016). https://doi.org/10.1038/srep20232
9. Tomasello, R., Garagnani, M., Wennekers, T., Pulvermller, F.: Brain connections of words, perceptions and actions: a neurobiological model of spatio-temporal semantic activation in the human cortex. Neuropsychol. **98**, 111–129 (2017)
10. Shrivastava, P., Bhoyar, K.K., Zadgaonkar, A.S.: Bridging the semantic gap with human perception based features for scene categorization. Int. J. Intell. Comput. Cybern. **10**(3), 387–406 (2017)
11. Colombino, T., Martin, D., Grasso, A., Marchesotti, L.: Reformulation of the semantic gap problem in content-based image retrieval scenarios. In: Lewkowicz, M. et al. (eds.) Proceedings of COOP 2010, Computer Supported Cooperative Work, Springer (2010)
12. Li, Y., Leung, C.H.C.: Multi-level semantic characterization and re-finement for web image search. Procedia Environ. Sci. **11**, 147–154 (2011). https://doi.org/10.1016/j.proenv.2011.12.023. (Elsevier Ltd.)
13. Li, X., Uricchio, T., Ballan, L., Bertini, M., M. Snoek, C.G., Bimbo, A.D.: Socializing the semantic gap: a comparative survey on image tag assignment, refinement, and retrieval. ACM Comput. Surv. **49**(1), 14 (2016)
14. Alzu'bi, A., Amira, A., Ramzan, N.: Semantic content-based image retrieval: a comprehensive study. J. Vis. Commun. Image Represent. **32**, 20–54 (2015)
15. Mesnil, G., Bordes, A., Weston, J., Chechik, G., Bengio, Y.: Learning semantic representations of objects and their parts. Mach. Learn. **94**(2), 281–301 (2014)
16. Singh, S., Sontakke, T.: An effective mechanism to neutralize the semantic gap in Content Based Image Retrieval (CBIR). Int. Arab J. Inf. Technol. **11**(2) (2014)
17. Montazer, G.A., Giveki, D.: Content based image retrieval system using clustered scale invariant feature transforms. Optik—Int. J. Light Electron Opt. **126**(18), 1695–1699 (2015)
18. Srivastava, P., Khare, A.: Integration of wavelet transform, Local Binary Patterns and moments for content-based image retrieval. J. Vis. Commun. Image Represent. **42**, 78–103 (2017)
19. Dong, H., Yu, S., Wu, C., Guo, Y.: Semantic Image Synthesis via Adversarial Learning. Accepted to ICCV 2017, Subjects: Computer Vision and Pattern Recognition (cs.CV), arXiv:1707.06873v1 [cs.CV] (2017)
20. Yasmin, M., Mohsin, S., Sharif, M.: Intelligent image retrieval techniques: a survey. J. Appl. Res. Technol. **12**(1), 87–103 (2014)
21. Khodaskar, A., Ladhake, S.: Semantic image analysis for intelligent image retrieval. Procedia Comput. Sci. **48**, 192–197 (2015)

Chapter 2
Low-Level Feature Detectors and Descriptors for Smart Image and Video Analysis: A Comparative Study

D. Avola, L. Cinque, G. L. Foresti, N. Martinel, D. Pannone and C. Piciarelli

Abstract Local feature detectors and descriptors (hereinafter extractors) play a key role in the modern computer vision. Their scope is to extract, from any image, a set of discriminative patterns (hereinafter keypoints) present on some parts of background and/or foreground elements of the image itself. A prerequisite of a wide range of practical applications (e.g., vehicle tracking, person re-identification) is the design and development of algorithms able to detect, recognize and track the same keypoints within a video sequence. Smart cameras can acquire images and videos of an interesting scenario according to different intrinsic (e.g., focus, iris) and extrinsic (e.g., pan, tilt, zoom) parameters. These parameters can make the recognition of a same keypoint between consecutive images a hard task when some critical factors such as scale, rotation and translation are present. The aim of this chapter is to provide a comparative study of the most used and popular low-level local feature extractors: SIFT, SURF, ORB, PHOG, WGCH, Haralick and A-KAZE. At first, the chapter starts by providing an overview of the different extractors referenced in a concrete case study to show their potentiality and usage. Afterwards, a comparison of the extractors is performed by considering the Freiburg-Berkeley Motion Segmentation (FBMS-59) dataset, a well-known video data collection widely used by the computer vision community. Starting from a default setting of the local feature extractors, the aim of the comparison is to discuss their behavior and robustness in terms of invariance with respect to the most important critical factors. The chapter also reports comparative considerations about one of the basic steps based on the feature extractors: the matching process. Finally, the chapter points out key considerations about the use of the discussed extractors in real application domains.

D. Avola · G. L. Foresti · N. Martinel · C. Piciarelli
Department of Mathematics, Computer Science and Physics, University of Udine,
Via Delle Scienze 206, 33100 Udine, Italy

L. Cinque · D. Pannone (✉)
Department of Computer Science, Sapienza University, Via Salaria 113,
00198 Rome, Italy
e-mail: pannone@di.uniroma1.it

© Springer International Publishing AG, part of Springer Nature 2018
H. Kwaśnicka and L. C. Jain (eds.), *Bridging the Semantic Gap in Image and Video Analysis*, Intelligent Systems Reference Library 145,
https://doi.org/10.1007/978-3-319-73891-8_2

2.1   Introduction

Nowadays, the computer vision is used in an increasing number of applications to support human activities in everyday life. A common aspect of these applications, that distinguishes them from simple monitoring systems in which human operators supervise video streams, is that they must have a certain degree of autonomy in understanding and interpreting the actions and events that occur within the video sequences. Currently, a wide range of smart applications are commonly used in different critical fields, including video surveillance [1–4], person re-identification [5–9], event detection [10–13] and others. The main step to implement any smart machine vision system, for any purpose, is to extract from the frames that compose a video stream a set of salient features (i.e., the above mentioned keypoints) through which to provide a significant abstraction of the background and foreground elements represented within the video. This abstraction, observed over time and processed by means of intelligent algorithms, is aimed to provide a semantic interpretation of what is happening in the video stream. In modern smart systems, this interpretation should always be done in real-time to ensure short response times for critical events, such as violence action detection, dangerous person identification, anomalous event recognition, and so on. To provide an overview of the main steps required during the design and running of a smart machine vision system, in Fig. 2.1 a generic architecture is reported.

As shown, such a system can be considered as consisting of four basic stages [14–16]. In the first, usually with the support of a dataset, a set of feature extractors are chosen and/or ad-hoc developed according to both the specific task and the image domain. Often, this stage can be redesigned several times to be reasonably certain that the identified extractors are sufficiently discriminative and robust. Subsequently, a classifier based on machine learning [17, 18] or deep learning [19, 20] techniques is used to define the different classes of actions or events that can be recognized by the system. When a system must recognize many complex classes, supervised techniques are preferred. By them, the classifier is first trained through a certain set of video sequences (learning phase) and then, by using another set of similar video

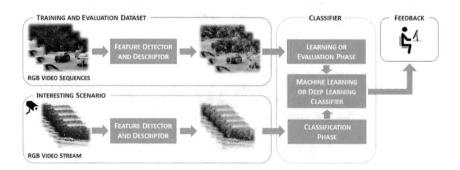

Fig. 2.1   General purpose pipeline of a smart machine vision system

sequences (evaluation phase), the recognition rate of the system is estimated. In other cases, instead, when a system must recognize few simple classes, unsupervised techniques can be also considered. By them, the classifier is immediately designed and parameterized to recognize a restricted set of events or actions. Usually, unsupervised techniques are used when the distinction of the few classes is truly apparent and very few feature extractors can be used to support the classification process. In these cases, only a dataset is needed and a wide number of video sequences (evaluation phase) is required to measure the classifier performance. Regardless of the specific used technique, if a classifier does not reach the estimated success rate, the whole process can be questioned, including the feature extractors. Once obtained a satisfactory classifier, the third stage treats the use of the system in an interesting scenario, and the last stage regards the automatic feedback to the users. Anyway, the pipeline has highlighted how the feature detectors and descriptors are at the base of each smart machine vision system.

Due to their importance, this chapter reports and compares a selected set of the most common and effective feature extractors currently well-known in literature. In particular, the chapter is focused on the following seven algorithms: SIFT [21], SURF [22], ORB [23], PHOG [24], WGCH [25], Haralick [26] and A-KAZE [27]. These extractors are chosen due to their use in a wide number of smart machine vision systems [28–31]. The comparison of the different extractors is performed by using the FBMS-59 [32], a popular public dataset widely used to train and evaluate different types of smart systems. During both the overview of the extractors (referenced by linked case studies) and the comparison among them, different key observations related to the critical aspects of the video sequence processing (i.e., intrinsic and extrinsic parameters) are reported and discussed.

The chapter is structured as follows. Section 2.2 presents a brief description of each low-level feature extractor. Moreover, some case studies are introduced and observations related to their behavior are also reported. Section 2.3 presents a comparison of the treated extractors. In particular, on the basis of both selected sample images and a default parametrization of each extractor, the main characteristics and limits of each extractor are discussed. In addition, since the matching process can be considered a basic step of each video processing based system, considerations about this topic are also highlighted. Finally, Sect. 2.4 concludes the chapter.

2.2 Low-Level Feature Detectors and Descriptors

This section is divided into two main sub-sections. The first provides a concise overview of SIFT, SURF, ORB, and A-KAZE. For each extractor, a reference work of the current state-of-the-art is reported and discussed. We have decided to treat separately these four extractors because they have often a key role in a wide range of smart machine vision systems. The rest of the extractors, i.e., PHOG, WGCH, and Haralick, are reported in the second sub-section. Unlike the first group of feature extractors, the latter are introduced and explained by means of a single reference work. This is due to the fact that these extractors often collaborate among them (or with other extractors) to support a specific smart machine vision system.

2.2.1 SIFT, SURF, ORB, and A-KAZE Extractors

These four extractors can be considered among the most powerful algorithms to identify robust keypoints in any kind of image. In each of them, the detection and description sub-algorithms are implemented separately to allow a more versatile use of the extractor. In fact, the separation allows to customize and/or improve, for a specific method, each main component of the extractor. In addition, the detector and/or the descriptor of an extractor can be used in combination with the components of another extractor to form hybrid approaches.

2.2.1.1 SIFT Algorithm

The Scale-Invariant Feature Transform (SIFT) algorithm [21] is designed to be robust to the most common critical factors in image and video analysis, including scale, rotation, and translation. In the practice, this extractor shows remarkable results also in presence of noise and illumination changes [33, 34]. The architecture of the SIFT algorithm is divided in the following four main stages, ranging from the detection of each point of interest up to their description [21]:

- Scale-Space Extrema Detection: By using a Difference-of-Gaussian (DoG) function, potential keypoints over all scales and image locations are searched;
- Keypoint Localization: By utilizing a reference model, each detected location is analyzed. A measure of stability to evaluate the candidate keypoint is used;
- Orientation Assignment: By making use of local image gradient directions, different orientations to each selected keypoint location are assigned;
- Keypoint Descriptor: By measuring local gradients at the selected scale around each keypoint, a significant pattern to characterize that specific area is computed.

More specifically, the Scale-Space Extrema Detection stage is used to identify the local extrema of an image by a linked multi-scale representation. This latter is computed by the convolution of the image, that is derived by a set of Gaussian Kernel with increasing variance (also named scale parameter). The subsampling and blurring of an image, to perform gradually the research of keypoints within the different version of it, is a robust and effective technique. In particular, given an image $I(x, y)$ and a Gaussian filter $G(x, y, \sigma)$, the scale-space, $L(x, y, \sigma)$, associated with the image can be represented as follows:

$$L(x, y, \sigma) = G(x, y, \sigma) * I(x, y) \qquad (2.1)$$

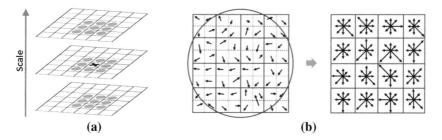

(a)   **(b)**

Fig. 2.2 SIFT algorithm: **a** detection of the local maximum and local minimum, **b** image gradients (left) and keypoint descriptor (right)

where, (x, y) is the generic point of the image, σ is the variance, $*$ is the convolution operator in x and y, and the Gaussian filter is expressed by the well-known equation:

$$G(x, y, \sigma) = \frac{1}{2\pi\sigma^2} e^{-\frac{(x^2+y^2)}{2\sigma^2}} \tag{2.2}$$

The construction of the scale-space is achieved by filtering iteratively the image at regular intervals, thus obtaining a set of processed images (named octaves). To detect the points of interest in the scale-space, the DoG function is computed. This latter is derived by calculating the difference of two nearby scales separated by a constant multiplicative factor k. Formally, it can be expressed as follows:

$$D(x, y, \sigma) = \big(G(x, y, k\sigma) - G(x, y, \sigma)\big) * I(x, y) = L(x, y, k\sigma) - L(x, y, \sigma) \tag{2.3}$$

The Keypoint Localization stage consists in determining the local extrema of the $D(x, y, \sigma)$ function, which represent the points of interest that must be identified. To determine them, each point is compared with its eight-neighbors in the current image and with its nine-neighbors in both upper and lower scale images (Fig. 2.2a). To be a local minimum (or a local maximum) a point must be the smaller (or the larger) in all the comparisons with respect to the other analyzed points.

Once computed the coordinates and scale of each point of interest, the Orientation Assignment stage consists in calculating the direction of each of them to ensure a high level of robustness with respect to the rotations. The scale associated with each point is used to choose the image with the closest scale. Subsequently, for each element of the chosen scale-space, $L(x, y)$, the module $m(x, y)$ and the orientation $\phi(x, y)$ of the local gradient can be quantified by applying the difference between pixels. The two functions can be formalized as follows:

$$m(x, y) = \sqrt{(L(x + 1, y) - L(x - 1, y))^2 + (L(x, y + 1) - L(x, y - 1))^2} \tag{2.4}$$

$$\varphi(x, y) = tan^{-1}\frac{L(x, y + 1) - L(x, y - 1)}{L(x + 1, y) - L(x - 1, y)} \tag{2.5}$$

At last, a histogram of the orientations is derived (with 36 bins that cover all the possible 360°). The peaks in the histogram correspond to the dominant orientations of the local gradients. In the final step of this stage, both the highest peak and all the peaks near to it (i.e., according to a fixed threshold) are identified, thus providing a single point of interest with a fixed direction.

Up to this stage, the SIFT algorithm has computed, for each keypoint, the spatial coordinates, the scale changes, and the orientation. In the last stage, Keypoint Descriptor, the algorithm assigns to each keypoint a descriptor that considers the different rotations of the image. To generate such a descriptor, a fixed area around each keypoint within the different scales of the image is selected. To ensure the invariance with respect to the rotations, the coordinates of each descriptor and the linked local gradient directions are rotated to cover all the possible rotation angles. Subsequently, the orientations of the samples that surround each keypoint are grouped into several sub-regions (typically, a grid of 4×4 sectors) and for each of them a histogram is computed (typically, with 8 bins). Each bin of the histogram corresponds to a different direction and covers 45° (Fig. 2.2b). Due to both the subdivision of the image in several zones and the linked computation of the histograms, the SIFT descriptor has a high dimensionality. This aspect can be considered a drawback of the SIFT algorithm especially for running online applications [22].

An interesting case study that uses SIFT features regards the development of the Facial Expression Recognition (FER) systems. These systems are designed to recognize, for different purposes, the main facial expressions of the human face. In Fig. 2.3 some facial expressions and the linked SIFT features are presented.

Usually, these features are adopted as first step to fix some strategic points on the human face. Subsequently, these points supported by other techniques are utilized to classify the facial expressions. A recent and interesting work that makes use of the SIFT features to implement a very robust FER system is reported in [35]. In

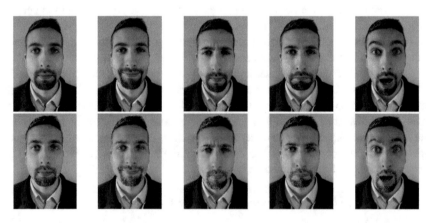

Fig. 2.3 SIFT features: the first five images (first row) show some well-known facial expressions: neutral, smiling, altered, doubtful, surprised. The other images (second row) show the extraction of the SIFT features on them

this paper, the authors propose a novel Deep Neural Network (DNN)-driven feature learning method and adopt it to implement a multi-view FER system. A fascinating aspect of this work is that the authors have been inspired by modern literature in neural cognition systems for the recognition of facial emotions. These studies state that the brain perception of facial expressions can be divided into several major periods happening in different brain areas [36]. The first period is about the low level salient image feature extraction occurring in the occipitotemporal cortex. The other periods are about the high level emotional semantic feature learning as well as the emotion perception happening in other brain areas (e.g., frontoparietal cortex, orbitofrontal cortex, amygdala). To emulate the first period, the authors detect those salient facial landmarks covering the main expression units of faces and subsequently extract low level SIFT descriptors from those salient facial landmarks as robust local appearance models to input the sequent network units. The other periods are emulated by introducing the projection and convolutional layers into the DNN. These layers are used to learn discriminative facial features across different facial landmark points and to extract high level features instead of 2D filters as in conventional Convolution Neural Networks (CNNs), respectively, thus obtaining results able to outperform the current state-of-the-art in this application field.

2.2.1.2   SURF Algorithm

As reported in the previous sub-section, the high dimensionality of the SIFT descriptor could be considered a critical aspect especially in those applications that require a substantial real-time processing [22]. Actually, the same research group that developed the SIFT algorithm tried to improve it by providing a best-bin-first alternative [21] to speed up the matching step, but they obtained results with a lower accuracy. All these reasons promoted the development of the Speeded Up Robust Feature (SURF) [22] algorithm. This algorithm has a quality comparable with that of SIFT as regards the management of the critical aspects (i.e., scale, rotation, translation, noise, and illumination changes), but it requires of a lower computational cost. As for the SIFT algorithm, also the architecture of SURF can be considered divided in four stages, ranging from the image reconstruction up to the description of each keypoint. In each stage, the SURF algorithm introduces an alternative approach with respect to the SIFT algorithm aimed to speed-up the processing [22]:

- Image Reconstruction: Unlike SIFT, the SURF algorithm utilizes, during the whole image processing, the integral images to reduce the computation time;
- Keypoint Localization: By adopting a Hessian matrix, the detection of the keypoints is obtained with a high accuracy and with a very low computation time;
- Orientation Assignment: By making use of a Haar-wavelet, this stage identifies a reproducible orientation for each keypoint within the image;
- Keypoint Descriptor: By constructing a square region centered around each keypoint and by using the previously orientations, a set of descriptors is obtained.

One of the main differences between SIFT and SURF is that this last uses integral images [37] in the whole image processing. Briefly, an integral image can be seen as a simplified version of an original image that allows for the fast implementation of box type convolution filters. Formally, it can be expressed as follows:

$$I_\Sigma(x, y) = \sum_{i=0}^{i \le x} \sum_{j=0}^{i \le y} I(x, y) \tag{2.6}$$

where, $I_\Sigma(x, y)$ is the integral image and $I(x, y)$ is the original image. The images obtained during the Image Reconstruction stage are used during the whole image processing to support all the other steps of the SURF algorithm.

In the Keypoint Localization, the algorithm utilizes a Hessian matrix, which is approximated by means of the use of the integral images, thus allowing high performance without losing in accuracy. The detector, so modified, assumes the name of Fast-Hessian. The SURF algorithm uses a Gaussian filter that allows spatial analysis and scale factors wider than SIFT algorithm. Formally, given a point $p = (x, y)$ in an image $I(x, y)$, the Hessian matrix, $H(p, \sigma)$, to the scale σ can be defined as follows:

$$H(p, \sigma) = \begin{bmatrix} L_{xx}(p, \sigma) & L_{xy}(p, \sigma) \\ L_{xy}(p, \sigma) & L_{yy}(p, \sigma) \end{bmatrix} \tag{2.7}$$

where, $L_{xx}(p, \sigma)$, $L_{yy}(p, \sigma)$, and $L_{xy}(p, \sigma)$, are the convolutions of the Gaussian second order derivatives according to the ∂_{xx}, ∂_{yy}, and ∂_{xy}, respectively.

In Orientation Assignment stage, for each keypoint a centered circular neighborhood with a fixed radius is calculated (at any scale). Then, within each area a Haar-wavelet in x and y direction is computed. The dominant direction is estimated by quantifying the sum of all the Haar-wavelet responses by means of a sliding window with a fixed size (usually $\pi/3$, as shown in Fig. 2.4a). The vector with the major module, among all those calculated, represents the main orientation of the keypoint.

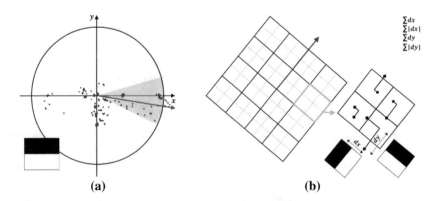

(a)   (b)

Fig. 2.4 SURF algorithm: **a** orientation calculation, within the sliding window (in azure) the responses of the Haar-wavelets are added up, **b** keypoint descriptor

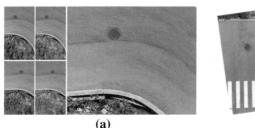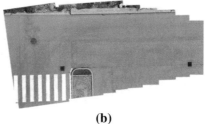

(a)   (b)

Fig. 2.5 SURF algorithm: **a** Running example of a mosaic composed by 2 frames. The first 2 frames (first column) are the original ones acquired by the UAV, the other 2 frames (second column) are the same frames in which the SURF features are extracted, the comprehensive view (third column) is the mosaic obtained by the stitching (based on the features), **b** example of mosaic composed by 9 frames

Once identified the orientation of each keypoint, the linked descriptor is implemented in the Keypoint Descriptor stage. In particular, the region is divided into a fixed number of sub-regions (usually, a grid of 4×4 sectors, as shown in Fig. 2.4b). For each of them, the Haar-wavelet response in dx and dy direction is computed. In order to increase the robustness with respect to the geometrical deformations, the responses of the wavelets are weighed with a Gaussian function centered in the point of interest. Finally, the different responses are summed among them (i.e., $\sum dx$ and $\sum dy$, see Fig. 2.4b), thus forming a first set of values linked to the descriptor. To keep also information about the polarity of the intensity changes, the absolute values of the obtained sums are also calculated (i.e., $\sum |dx|$ and $\sum |dy|$, see Fig. 2.4b).

A particular case study that uses SURF features regards the development of Unmanned Aerial Vehicle (UAV) based systems. These systems are designed to accomplish a very wide range of tasks, including object recognition, vehicle tracking, land monitoring, and others. A key requirement of many of these tasks is the mosaic construction of a specific area of interest. A mosaic is a comprehensive view of a target area obtained by stitching the frames acquired by a UAV that flies over it. In Fig. 2.5 two examples (with 2 frames and 9 frames, respectively) are provided.

In this context, an interesting work is reported in [38], where the authors propose an efficient system for mosaicking wide areas by using SURF features. The authors address the problem of video surveillance in wide outdoor environments in which common technologies (e.g., fixed PTZ cameras) are not suitable or sufficient.

2.2.1.3 ORB Algorithm

As previously reported, an extractor is composed of two main components: detector and descriptor. The Oriented FAST and Rotated BRIEF (ORB) [23] algorithm is based on the combination of two well-known methods. The first, FAST, is used as detector, while the second, BRIEF, is used as descriptor. The ORB algorithm is designed to be an efficient alternative to SIFT and SURF.

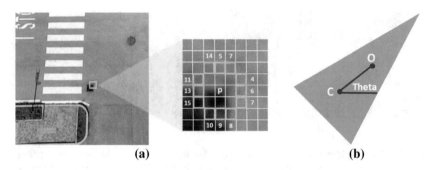

Fig. 2.6 ORB algorithm: **a** FAST detector, **b** orientation of the keypoint

The FAST method analyzes the brightness of the image around to each potential keypoint. In particular, if a surrounding circular area centered in a point has an arc of contiguous pixels with a suitable length (usually, 3/4 of the perimeter of the circle) and a substantial illumination change, then that point is considered a point of interest. An intuitive example is shown in Fig. 2.6a. In a first step, the FAST method uses an expedient to speed-up the processing. It consists in analyzing only 4 points on the circle with an offset of 90°. If at least three of these points do not have a correlated brightness with respect to a fixed threshold, then the potential point is discarded immediately. The BRIEF method is based on the idea that image patches can be effectively classified on the basis of a relatively small number of pairwise intensity comparisons. Since, the original FAST method does not provide the orientation component, a variation of the FAST method (named oFAST) is computed. For each keypoint oFAST computes the vector, i.e., the orientation, from the center O up to the centroid C within the patch that contains the corner (see Fig. 2.6b).

An interesting case study that uses ORB features regards the development of the Situation AWareness (SAW) systems. These systems are designed to interpret the events that occur within a video stream. Since, usually, these systems are adopted to support surveillance tasks, the role of the real-time performance is evident. In [39], the authors propose a method to detect moving objects in moving background. As well-known, this is a very complex task in computer vision because both the movements of the camera must be continuously estimated to model and update the background and the foreground objects must be continuously separated from the background in order to track them. Despite this, the authors provide a robust and real-time system that takes full advantage from the main intrinsic features of ORB.

2.2.1.4   A-KAZE Algorithm

The Accelerated-KAZE (A-KAZE) algorithm [27] is one of the most recent feature extractors. It is having a remarkable consideration by the computer vision community due to its successful use in a growing number of smart systems. It is an evolution of the KAZE [40] algorithm. As shown in the previous sub-sections, almost all the feature

(a)   **(b)**

Fig. 2.7 A-KAZE algorithm: **a** DFRobotShop Rover V2, **b** an example of test in uncontrolled environment to search damages in a pipe by using A-KAZE features

extractors adopt the Gaussian kernel to produce the scale-space representation of an original image $I(x, y)$. This approach, on one side, supports the noise reduction and emphasizes the prominent structures of the image, on the other hand, it presents some important drawbacks. In fact, Gaussian blurring does not respect the boundaries of objects and smoothest to the same degree both details and noise at all scale levels. All these factors produce a significant reduction of the localization accuracy [40]. To overcome all these issues, the KAZE algorithm adopts a nonlinear scale-space by using Additive Operator Splitting (AOS) techniques and variable conductance diffusion. The main contribution of the algorithm presented in [27] to that shown in [40] is the introduction of the Fast Explicit Diffusion (FED) to dramatically speed-up feature detection in nonlinear scale spaces. At the current state-of-the-art, and considering the remarkable results shown by the authors, the A-KAZE algorithm can be considered one of the best feature extractors. In Fig. 2.7, a recent example of the use of the A-KAZE features is reported.

One of the most fascinating fields of all time is robotics. The authors of the work proposed in [41] propose a multipurpose autonomous robot for target recognition in unknown environments. Inside the robot are implemented both a simultaneous localization and mapping (SLAM) algorithm and an object identification algorithm fully based on the A-KAZE features. In this specific context, the ability of the A-KAZE algorithm to speed-up the feature extraction and matching processes has played a key role due to the necessity to perform the research task in real-time.

2.2.2 PHOG, WGCH, and Haralick Extractors

These three extractors can be considered a suitable choice to observe different ways, with respect to the previous ones, to extract features from an image. Actually, the three algorithms treated in this sub-section cannot be considered "pure" extractors since they implement only the description stage. The detection stage is created ad-hoc according to the specific case study. Anyway, a common practice is the use of detectors coming from other approaches to support the reported three descriptors.

2.2.2.1 Running Example Focused on Person Re-Identification

In the last years, one of the most promising research field in computer vision is represented by target re-identification in distributed wide camera networks [42]. The problem of re-identifying targets moving across cameras with non-overlapping fields of view (FoVs) is challenging due to the open issues of multi-camera video analysis, such as changes of scale, illumination, viewing angle and pose. The task is even harder when dealing with people due to the non-rigid shape of the human body. To address these issues and build a discriminating signature, four local and global features are extracted and accumulated over multiple frames. As shown in Fig. 2.8, different features are usually extracted: (i) Pyramid Histogram of Orientation Gradients (PHOG) [24]; (ii) SIFT [21] (Sect. 2.2.1.1); (iii) SIFT based weighted Gaussian color histogram (WGCH) [25]; (iv) Haralick texture features [26].

Each of these features has been properly selected to capture different information about the given image. PHOG features capture the shape and the spatial layout of the person silhouette. SIFT and WGCH features capture the appearance of the person at specific local regions of interest. Finally, Haralick features capture information about textures. The process of extracting such features is described in detail in the following. The SIFT features have been already described previously (Sect. 2.2.1.1)

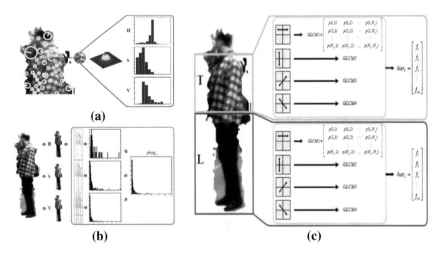

Fig. 2.8 Computed features. **a** SIFT-based Weighted Gaussian Color Histograms (observe that, SIFT is used only as detector of keypoints, which are described by WGCH), **b** PHOG features, **c** Haralick features for the two detected body parts

no further information is given. They are introduced in this case study to supply a detector stage to the WGCH features (i.e., description stage).

PHOG Algorithm

The PHOG feature [24] captures the local shape and the spatial layout of shape in a given image [43]. To this end the spatial pyramid framework is exploited [44]. In the latter, the given image is divided into a sequence of spatial grid cells by repeatedly doubling the number of divisions in each axis direction. That is, the number of points in a grid cell at one level is the sum of the points contained in the four cells it is divided into at the next level of the pyramid. The number of grid cells at each level of the pyramid gives the number of HOGs that must be computed at that level. The PHOG feature vector is computed as a concatenation of all the HOG vectors computed at all the grid cells locations at each level of the spatial pyramid representation, where each bin in the local HOG feature represents the number of edge gradients that have orientations within a certain oriented (i.e., angular) range.

The contribution of each gradient to the histogram is weighted by the magnitude of the gradient itself and, similarly to SIFT feature computation, a soft assignment is used to affect neighboring bins. More formally, let K be the number of orientations bins used to compute a single HOG feature vector, and $l = 0, 1, \ldots, L$ be the level of the spatial pyramid representation such that the number of grid cells at level l of the spatial pyramid is $2l$ along each dimension, e.g., at level 0, the concatenated HOG feature vector is of size K. Let HOG_K^l be the concatenation of the HOG feature vectors computed for all the $4l$ grid cells. Then the PHOG feature vector for the entire image is a column vector of length $m = K \sum_l^L 4^l$ computed as:

$$phog = [HOG_K^0, HOG_K^1, HOG_K^2, \ldots, HOG_K^l, \ldots, HOG_K^L] \qquad (2.8)$$

The PHOG feature vector is finally normalized to sum-up to unity. Figure 2.9 illustrates this principle showing the PHOG features computed for different number of levels L. In the implementation provided in this running example, before extracting the PHOG features from the whole silhouette, $\hat{F}(L)$ is histogram equalized and projected into the HSV color space to achieve illumination invariance. As shown in Fig. 2.9, to retain some information about colors, the gradients for each of the hue, saturation and value axes are computed separately only at image locations where an edge is detected by the Canny edge detection algorithm. The PHOG feature matrix $PHOG \in R^{mx3}$, computed for the given image I is defined as:

$$PHOG(L) = [phog_h, phog_s, phog_v] \qquad (2.9)$$

where, $phog_h$, $phog_s$, and $phog_v$ are the phog feature vectors computed for the hue, saturation and value color components respectively.

SIFT and WGCH Algorithm

The SIFT features are jointly used with the WGCH features to capture the local chromatic appearance of given person image. Given the silhouette of the whole body

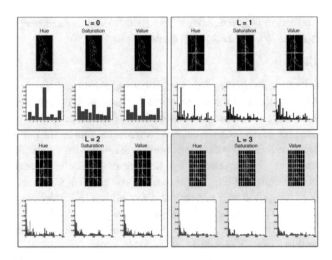

Fig. 2.9 Effects of the number of levels (L) used to compute the PHOG feature. PHOG features extracted from the hue, saturation and value color components using different spatial pyramid levels are shown. For each of the four blocks, the top row shows the grid cells (in green) at which the HOG features are extracted. Bottom rows show the final PHOG features computed concatenating the HOG features extracted at each level of the pyramid

$\hat{F}(L)$, the SIFT features are computed. Then, for each of the detected SIFT features a circular image patch centered at the SIFT keypoint is extracted. The three-color axes that compose it are separately taken to compute three different histograms weighted by a Gaussian distribution. Due to the robust identification of localized SIFT keypoints, and to the fact that the farthest part of the patch is given a lower weight, the WGCH captures the local chromatic appearance reducing the occlusion and viewpoint changes issues. Let define a single SIFT feature as:

$$sift = [sift_{kp}, sift_{hist}, sift_F] \tag{2.10}$$

where $sift_{kp} = [x, y]^T$ gives the x and y coordinates of the detected keypoint, $sift_{hist} \in R^{128}$ is the standard SIFT feature descriptor and $sift_F \in \{T, L\}$ denotes the body part region from which the feature is extracted. All the detected SIFT features are then concatenated to form a larger feature vector:

$$SIFT(L) = [sift(1), sift(2), \ldots, sift(S)] \tag{2.11}$$

where $sift(k)$ is the k-th SIFT feature extracted from the silhouette $\hat{F}(L)$. Given a SIFT feature keypoint $sift_{kp}$, the process of computing the related WGCH feature is shown in Fig. 2.10.

A circular patch R of radius r centered at $sift_{kp}$ is extracted and projected into the HSV color space to better cope with illumination changes and color variations. To compute the WGCH feature vector, here denoted as $wgch$, each element of the

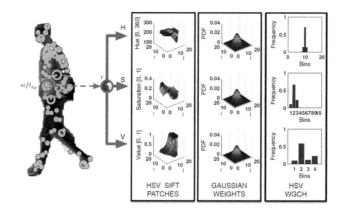

Fig. 2.10 WGCH. The process of computing the WGCH related to a specific SIFT keypoint $sift_{kp}$ is shown. A circular patch of radius r centered at $sift_{kp}$ is extracted and projected to the HSV color space. First column shows the hue, saturation and value intensities of the given patch. Second column shows the Gaussian weights used to weight the HSV sift patches values. Third column shows the three WGCHs computed for the hue, saturation and value axes using different bin quantizations

patch R at coordinates (i, j) is weighted by the probability density value at (i, j) of a Gaussian probability density function with mean $\mu = [r/2, r/2]$ and diagonal covariance $\Sigma \in R^{2x2}$. This can be written as follows. Let $[b, t)$ be a single bin range of the WGCH and $R_{i,j}$ be the pixel value at coordinates (i, j) of the patch R, then, if $b \leq R_{i,j} < t$:

$$wgch(b, t) = wgch(b, t) + N(\mu, \sum)_{i,j} \tag{2.12}$$

where $N(\mu, \sigma)_{i,j}$ is the value at location (i, j) of a Gaussian probability density function. The computed WGCH is then normalized to sum up to 1. Since the WGCH is computed for the hue, saturation and value patches, we end up with three WGCHs denoted as $wgch_h \in R^{b_h}$, $wgch_s \in R^{b_s}$, and $Wgch_v \in R^{b_v}$ where b_n, b_s, and b_v are the number of bins used for quantization of the hue, saturation and value components respectively. A single WGCH feature is denoted as:

$$wcgh = [wcgh_h, wcgh_s, wcgh_v] \tag{2.13}$$

As WGCH features are extracted from the previously computed SIFT features. Finally, the WGCH feature matrix can be defined as follows:

$$WCGH(L) = [wcgh(1), wcgh(2) \dots wcgh(S)] \tag{2.14}$$

where $wcgh(k)$ is the WGCH associated to the k-th SIFT feature $sift(k)$.

Haralick Algorithm

The Haralick feature captures information about the patterns that emerge in the image texture. In particular, Haralick feature captures information about the image textures

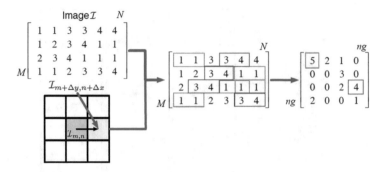

Fig. 2.11 GLCM. Given a gray scale input image I and a adjacency mask (bottom left) the gray level co-occurrence matrix (rightmost) is formed by counting the number of adjacent pixels that have gray intensity level equals to (a, b). Here an example of computing a GLCM using offset $\Delta x = 1$ and $\Delta y = 0$ and $ng = 4$ gray levels is shown. Green boxes show pixels with intensity values $(a = 1, b = 1)$ that are adjacent according to the offset. Red boxes highlight pixels with intensity values $(a = 3, b = 4)$ for the same offset

such as the homogeneity, the gray level linear dependencies, the contrast, the number and the nature of edges, and the complexity of the image itself. These features are calculated in the spatial domain, and they rely on the assumption that the texture information in an image is contained in the spatial relationship among the image gray levels. To extract the Haralick texture features, a gray level co-occurrence matrix (GLCM) is used. Such a matrix defined over an image describes the distribution of co-occurring gray level pixel values at a given offset. The gray-level co-occurrence matrix is a function of the angular relationship between the neighboring pixels in the image as well as a function of the distance between them.

As shown in Fig. 2.11, given an image L of size $M \times N$, with $a, b = 1, 2, \ldots, ng$ gray levels, the gray level co-occurrence matrix $GLCM \in R^{ng \times ng}$ defined over L is parameterized by the adjacency matrix (offsets Δx and Δy). Given such offset values, the GLCM is computed as:

$$GLCM_{a,b}^{\Delta x, \Delta y} = \sum_{n=1}^{N} \sum_{m=1}^{M} \begin{cases} 1, & if \ L_{m,n} == a \wedge L_{m+\Delta x, n+\Delta x == b} \\ 0, & otherwise \end{cases} \quad (2.15)$$

where $L_{m,n}$ is the gray level pixel intensity of image L at coordinates (m, n). Haralick features rely on the assumption that image texture information is contained in the GLCM, so Haralick features are extracted from the computed GLCM. However, the parameters Δx and Δy lead to different GLCM and different values for the pixel intensity pairs $(a.b)$ and (b, a). This would make the GLCM, hence the Haralick features, sensitive to rotation. To deal with this issue, the following suggestions can be considered [45]:

    i. use the following offset Δx and Δy values: $\Delta x = 1$, $\Delta y = 0$ (0°); $\Delta x = 1$, $\Delta y = 1$ (45°); $\Delta x = 0$, $\Delta y = 1$ (90°); $\Delta x = -1$, $\Delta y = 1$ (135°);

ii. the GLCM matrix entries as symmetric so that both $(a.b)$ and (b, a) pairings are computed by counting the number of times the value a is adjacent to the value b;

iii. advantage of pooling and average the resulting GLCM over multiple images.

If so, some invariance to rotations is achieved. To use Haralick features to compute a discriminative signature for re-identification, the assumption that most of the people wear different clothes for the bottom and for the lower body parts is considered. In light of this, two Haralick texture feature vectors are extracted: one for the torso and one for the legs silhouettes, respectively. Let $GLCM_T(L)$ and $GLCM_L(L)$ be the GLCM matrices computed for the torso and legs regions of a given person's images L. Then, following the details in [45], those are used to extract the two 14 dimensional feature vectors $HART_T(L) \in R^{14}$ and $HART_L(L) \in R^{14}$, where $HART_T(L)$ is the Haralick feature vector computed for the torso region and $HART_L(L)$ is the Haralick feature vector computed for the legs region.

2.3 Low-Level Feature Comparison and Discussion

This section is divided into two main sub-sections. The first analyzes the behavior and robustness of SIFT, SURF, ORB, A-KAZE, PHOG, WGCH, and Haralick in terms of invariance with respect to the most well-known critical aspects, including scale, rotation, and translation. By using each operator, the second sub-section, provides a concise overview about the matching process between two images that have an overlapped area. All the images used to produce the comparative observations come from the FBMS-59 dataset [46]. This last is a well-known video data collection widely used by the computer vision community. The FBMS-59 dataset consists of 59 challenging video sequences, for a total of 720 annotated frames. To support machine learning and deep learning techniques, the dataset is divided into a training set and a test set. The first sub-section presents three reference examples. In each of which an original image and the linked features extracted by the discussed extractors are reported. The examples have been chosen to cover a wide range of situations that can occur during the implementation of different types of smart systems. In the second sub-section, for each discussed feature extractor, key observations about the matching process are provided.

2.3.1 Behavior and Robustness

In Fig. 2.12, the application of the described feature extractors on a first reference image coming from the FBMS-59 dataset is reported.

The first chosen reference case regards an image of a car crossing a road. This is a good testing scene due to several factors, such as the sun reflection on the car,

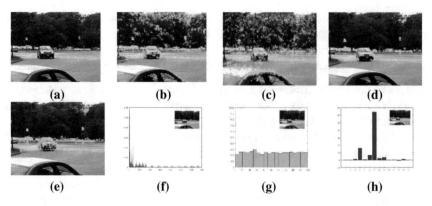

Fig. 2.12 Application to the **a** first reference image of **b** SIFT, **c** SURF, **d** ORB, **e** A-KAZE, **f** PHOG, **g** WGCH and **h** Haralick feature extractors

the multitude of details given by a parked car, and the trees on the background. As it is possible to observe, SIFT (Fig. 2.12b) and SURF (Fig. 2.12c) acts quite similar in finding keypoints. Keypoints are found within the same image areas, but due to their different implementation SURF extracts a greater number of keypoints. Moving towards binary descriptors, we have that also ORB (Fig. 2.12d) and A-KAZE (Fig. 2.12e) find keypoints in the same areas of image, and as for SIFT and SURF, their different implementation allows A-KAZE to extract more robust keypoints due to the M-LDB descriptor. A common property among these extractors is that on uniform surfaces, few or none keypoints are found. In this example, SURF is the extractor that finds most keypoint on the asphalt, while with ORB and A-KAZE no keypoints are found. By analyzing histograms approaches, PHOG (Fig. 2.12f) and WGCH (Fig. 2.12g) acts differently one from the other. This is since in PHOG the histograms are computed within cells equally spaced among the image, while the WGCH computes the histogram around a SIFT keypoint. Finally, texture approaches are analyzed. In our case, all the 14 Haralick (Fig. 2.12h) features are applied. In this reference case, the variance, the sum variance and the entropy are the most representative features, due to the non-uniformity of the image.

In Fig. 2.13, the application of the described feature extractors on a second reference image coming from the FBMS-59 dataset is reported.

The second example case chosen regards the image of a horse in a field. Differently from the first example, here we have a clear separation between foreground and background. In detail, since the background is a meadow it presents a more homogeneous distribution of the color, that could be challenging for binary features extractors. Concerning SIFT (Fig. 2.13b) and SURF (Fig. 2.13c), as the first example their keypoints distribution is similar within the image and, also in this example, SURF extracts more keypoints. Also for ORB (Fig. 2.13d) and A-KAZE (Fig. 2.13e) there is a similar keypoints distribution within the image, but A-KAZE seems performing better since it is able to extract some keypoints from the horse. Differently from the first example, where only SIFT and SURF extracted some keypoints from

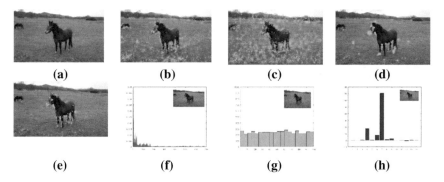

Fig. 2.13 Application to the **a** second reference image of **b** SIFT, **c** SURF, **d** ORB, **e** A-KAZE, **f** PHOG, **g** WGCH and **h** Haralick feature extractors

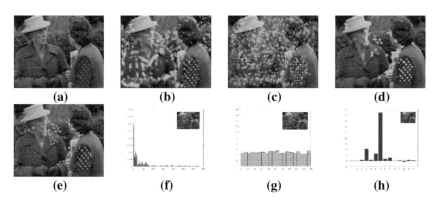

Fig. 2.14 Application to the **a** third example image of **b** SIFT, **c** SURF, **d** ORB, **e** A-KAZE, **f** PHOG, **g** WGCH and **h** Haralick feature extractors

the asphalt, here also ORB and A-KAZE could detect some keypoints. This is due to the blades of grass, which cast their shadow on the ground. Regarding PHOG (Fig. 2.13f) and WGCH (Fig. 2.13g), the first seems to perform better. This may be to the fact that the canny operator used by PHOG helps in extracting the shape of the horse. Finally, also in this example, all the 14 Haralick features are used. Due to the high homogeneity of the background, the entropy value for Haralick feature is less than the first example while other features have, within a small range, the same value obtained in the first example.

In Fig. 2.14, the application of the described feature extractors on a third reference image coming from the FBMS-59 dataset is reported.

The third and last example case regard an image portraying two talking women. This image has been chosen has third example due to the high number of details (i.e., colors and shaper) it provides. As recognized in the previous two examples, SIFT (Fig. 2.14b) and SURF (Fig. 2.14c) extract features in a similar distribution. In this last case, SURF keypoints have a more homogeneous distribution within the

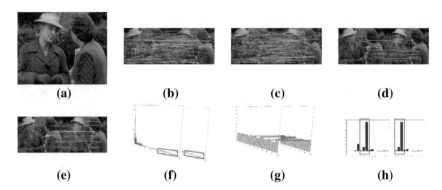

Fig. 2.15 Matching between the image of the third reference example with the image shown in **a** by using **b** SIFT, **c** SURF, **d** ORB, **e** A-KAZE, **f** PHOG, **g** WGCH and **h** Haralick keypoints

image, while SURF extract keypoints from well-defined areas, such as the collar of the brown jacket. Concerning ORB (Fig. 2.14d) and A-KAZE (Fig. 2.14e), the same keypoints are extracted. In detail, the only keypoints that are extracted with A-KAZE but not with ORB are the ones at the left of the head of the woman wearing the brown jacket. Differently from SIFT and SURF, ORB and A-KAZE have not extracted any keypoint from the red jacket worn by the woman on the right. Some relevant details, as the white dots on the shirt of the woman on the right, are detected by all the extractors. Concerning PHOG (Fig. 2.3f), there is a high concentration of the values on the lower bins as for the previous example. WGCH (Fig. 2.3g), instead, has increased its performance due to the high distribution of keypoints on relevant details of the image (e.g., on the women). Again, all the 14 Haralick features are computed also for the third example. As it is possible to observe, the obtained histogram (Fig. 2.3h) is similar to the one of the first example. This is due to the fact that in both examples the used images contains a high number of details, differently from the second example in which there is a uniform background.

2.3.2 Matching Process

In this sub-section, the matching process between two frames by using the described feature extractors is shown. An overview is provided in Fig. 2.15.

In order to analyze only the true matching keypoints between the two images, a ratio test has been performed as suggested in [21]. Since removing false matches is a hard task, there still are some false matches in the proposed images. To test the robustness to rotation and translation of the discussed feature extractors, the Fig. 2.15a has been chosen due to the head rotation of the woman at the left side. Due to the repeatability of the keypoints, we have that all the keypoints residing on the head of the woman that are in common between Figs. 2.14a and 2.15a are both translated and rotated. As it is possible to see, SIFT (Fig. 2.15b) and SURF (Fig. 2.15c) are the extraction with the highest number of matches. Unfortunately, with a high number of matches the probability of having a mismatch increases. In the proposed images, the mismatches are the ones in which the connecting line between keypoints ranging from top to bottom. Concerning ORB (Fig. 2.15d) and A-KAZE (Fig. 2.15e) there are less matches and, consequently, less mismatches. Differently from SIFT, SURF, ORB and A-KAZE, which features matching can be performed with algorithms such as brute-force and k-nearest-neighborhood, PHOG (Fig. 2.15f), WGCH (Fig. 2.15g) and Haralick (Fig. 2.15h) keypoints can be matched by using shape-matching and histograms comparison algorithms.

2.4 Conclusions

In this chapter, a comparative study of the most used and popular low-level local feature extractors (SIFT, SURF, ORB, PHOG, WGCH, Haralick and A-KAZE) is provided. In the first Section, a reference architecture for the smart systems and an overview of the related main components are presented. In the second Section, an overview of each feature extractor and a linked case study is reported. The Section has been focused on highlighting the main theoretical differences among the different extractors. In the third Section, key observations about the matching process are delineated. All the reference examples in this last section have been performed by using set of images coming from the public FBMS-59 dataset.

Acknowledgements This work was partially supported by both the "Proactive Vision for advanced UAV systems for the protection of mobile units, control of territory and environmental prevention (SUPReME)" FVG L.R. 20/2015 project and the "Augmented Reality for Mobile Applications: advanced visualization of points of interest in touristic areas and intelligent recognition of people and vehicles in complex areas (RA^2M)".

References

1. Xu, D., Yan, Y., Ricci, E., Sebe, N.: Detecting anomalous events in videos by learning deep representations of appearance and motion. Comput. Vis. Image Underst. **156**, 117–127 (2017)
2. Kulkarni, K., Turaga, P.: Reconstruction-free action inference from compressive imagers. IEEE Trans. Pattern Anal. Mach. Intell. **38**(4), 772–784 (2016)
3. Micheloni, C., Snidaro, L., Foresti, G.L.: Exploiting temporal statistics for events analysis and understanding. Image Vis. Comput. **27**(10), 1459–1469 (2009)
4. Zhang, T., Jia, W., He, X., Yang, J.: Discriminative dictionary learning with motion weber local descriptor for violence detection. IEEE Trans. Circuits Syst. Video Technol. **27**(3), 696–709 (2017)
5. Martinel, N., Micheloni, C., Foresti, G.L.: Kernelized saliency-based person re-identification through multiple metric learning. IEEE Trans. Image Process. **24**(12), 5645–5658 (2015)
6. Zhao, R., Oyang, W., Wang, X.: Person re-identification by saliency learning. IEEE Trans. Pattern Anal. Mach. Intell. **39**(2), 356–370 (2017)

7. Wang, T., Gong, S., Zhu, X., Wang, S.: Person re-identification by discriminative selection in video ranking. IEEE Trans. Pattern Anal. Mach. Intell. **38**(12), 2501–2514 (2016)
8. García, J., Martinel, N., Gardel, A., Bravo, I., Foresti, G.L., Micheloni, C.: Discriminant context information analysis for post-ranking person re-identification. IEEE Trans. Image Process. **26**(4), 1650–1665 (2017)
9. Tao, D., Guo, Y., Song, M., Li, Y., Yu, Z., Tang, Y.Y.: Person re-identification by dual-regularized kiss metric learning. IEEE Trans. Image Process. **25**(6), 2726–2738 (2016)
10. Piciarelli, C., Foresti, G.L.: Surveillance-oriented event detection in video streams. IEEE Intell. Syst. **26**(3), 32–41 (2011)
11. Bin, Y., Yang, Y., Shen, F., Xu, X.: Combining multi-representation for multimedia event detection using co-training. Neurocomputing **217**, 11–18 (2016)
12. Xian, Y., Rong, X., Yang, X., Tian, Y.: Evaluation of low-level features for real-world surveillance event detection. IEEE Trans. Circuits Syst. Video Technol. **27**(3), 624–634 (2017)
13. Wu, J., Hu, D.: Learning effective event models to recognize a large number of human actions. IEEE Trans. Multimedia **16**(1), 147–158 (2014)
14. Avola, D., Foresti, G.L., Piciarelli, C., Vernier, M., Cinque, L.: Mobile applications for automatic object recognition. Encycl. Inf. Sci. Technol. (Fourth Edition) **18**, 100–110 (2017)
15. Li, G., Yu, Y.: Visual saliency detection based on multiscale deep CNN features. IEEE Trans. Image Process. **25**(11), 5012–5024 (2016)
16. Avola, D., Cinque, L., Foresti, G.L., Massaroni, C., Pannone, D.: A keypoint-based method for background modeling and foreground detection using a PTZ camera. Pattern Recogn. Lett. **7**, 1–10 (2016)
17. Bishop, C.M.: Pattern Recognition and Machine Learning. In Series: Information Science and Statistics, 1st edn, 738 p. Springer (2007)
18. Camastra, F., Vinciarelli, A.: Machine Learning for Audio, Image and Video Analysis: Theory and Applications. In Series: Advanced Information and Knowledge Processing, 2nd edn, 580 p. Springer (2016)
19. Nath, V., Levinson, S.E.: Autonomous Robotics and Deep Learning. In Series: SpringerBriefs in Computer Science, 1st edn, 66 p. Springer (2014)
20. Goodfellow, I., Bengio, Y., Courville, A.: Deep Learning. In Series: Adaptive Computation and Machine Learning, 1st edn, 800 p. MIT Press (2016)
21. Lowe, D.G.: Distinctive image features from scale-invariant keypoints. Int. J. Comput. Vis. **60**(2), 91–110 (2004)
22. Bay, H., Ess, A., Tuytelaars, T., Gool, L.V.: Speeded-up robust features (SURF). Comput. Vis. Image Underst. **110**(3), 346–359 (2008)
23. Rublee, E., Rabaud, V., Konolige, K., Bradski, G.: ORB: an efficient alternative to SIFT or SURF. In: Proceedings of the IEEE International Conference on Computer Vision (ICCV), pp. 2564–2571. IEEE Computer Society (2011)
24. Bosch, A. Zisserman, A., Munoz, X.: Representing shape with a spatial pyramid kernel. In: Proceedings of the 6th ACM International Conference on Image and Video Retrieval (CIVR), ACM, pp. 401–408 (2007)
25. Wang, W., Zhang, J., Li, L., Wang, Z., Li, J., Zhao, J.: Image haze removal algorithm for transmission lines based on weighted Gaussian PDF. In: 6th International Conference on Graphic and Image Processing (ICGIP), Proc. SPIE-9443, pp. 1–7 (2015)
26. Haralick, R.M., Shanmugam, K., Dinstein, I.: Textural features for image classification. IEEE Trans. Syst. Man Cybern. **SMC-3**(6), 610–621 (1973)
27. Alcantarilla, P., Nuevo, J., Bartoli, A.: Fast explicit diffusion for accelerated features in nonlinear scale spaces. In: Proceedings of the British Machine Vision Conference, pp. 1–11. BMVA Press (2013)
28. Turaga, P., Chellappa, R., Subrahmanian, V.S., Udrea, O.: Machine recognition of human activities: a survey. IEEE Trans. Circuits Syst. Video Technol. **18**(11), 1473–1488 (2008)
29. Liu, H., Chen, S., Kubota, N.: Intelligent video systems and analytics: a survey. IEEE Trans. Industr. Inf. **9**(3), 1222–1233 (2013)

30. Krig, S.: Interest point detector and feature descriptor survey. In: Computer Vision Metrics: Survey, Taxonomy, and Analysis, 1st edn, pp. 217–282. Springer (2014)
31. Rodríguez, N.D., Cuéllar, M.P., Lilius, J., Calvo-Flores, M.D.: A survey on ontologies for human behavior recognition. ACM Comput. Surv. 46(4), 1–33 (2014)
32. Ochs, P., Malik, J., Brox, T.: Segmentation of moving objects by long term video analysis. IEEE Trans. Pattern Anal. Mach. Intell. 36(6), 1187–1200 (2014)
33. Lowe, D.G.: Object recognition from local scale-invariant features. In: Proceedings of the IEEE International Conference on Computer Vision (ICCV), pp. 1150–1157. IEEE Computer Society (1999)
34. Hannane, R., Elboushaki, A., Afdel, K., Naghabhushan, P., Javed, M.: An efficient method for video shot boundary detection and keyframe extraction using SIFT-point distribution histogram. Int. J. Multimedia Inf. Retr. 5(2), 89–104 (2016)
35. Zhang, T., Zheng, W., Cui, Z., Zong, Y., Yan, J., Yan, K.: A deep neural network-driven feature learning method for multi-view facial expression recognition. IEEE Trans. Multimedia 18(12), 2528–2536 (2016)
36. Adolphs, R.: Neural systems for recognizing emotion. Curr. Opin. Neurobiol. 12(2), 169–177 (2002)
37. Viola, P., Jones, M.: Rapid object detection using a boosted cascade of simple features. In: Proceedings of the IEEE Conference on Computer Vision and Pattern Recognition (CVPR), pp. I-511–I-518. IEEE computer Society (2001)
38. Piciarelli, C., Micheloni, C., Martinel, N., Vernier, M., Foresti, G.L.: Outdoor environment monitoring with unmanned aerial vehicles. In: Proceeding of the International Conference on Image Analysis and Processing (ICIAP), pp. 279–287. Springer (2013)
39. Ramya, R., Sudhakara, B.: Motion detection in moving background using ORB features matching and affine transform. Int. J. Innovative Technol. Res. 1(1), 162–164 (2015)
40. Alcantarilla, P., Bartoli, A., Davison, A.J.: KAZE Features. In: Proceeding of the European Conference on Computer Vision (ECCV), pp. 214–227. Springer (2012)
41. Avola, D., Foresti, G.L., Cinque, L., Massaroni, C., Vitale, G., Lombardi, L.: A multipurpose autonomous robot for target recognition in unknown environments. In: International Conference on Industrial Informatics (INDIN), pp. 766–771. IEEE Computer Society (2016)
42. Martinel, N., Foresti, G.L., Micheloni, C.: Person reidentification in a distributed camera network framework. IEEE Trans. Cybern. PP(99), 1–12 (2016)
43. Bosch, A., Zisserman, A., Munoz, X.: Representing shape with a spatial pyramid kernel. IN: Proceedings of 6th ACM International Conference on Image and video retrieval (CIVR), pp. 401–408. ACM Press (2007)
44. Lazebnik, S., Schmid, C., Ponce, J.: Beyond bags of features: spatial pyramid matching for recognizing natural scene categories. In: Proceedings of the International Conference on Computer Vision and Pattern Recognition (CVPR), pp. 2169–2178. IEEE computer Society (2006)
45. Haralick, R.M., Shanmugam, K., Dinstein, I.: Textural features for Image classification. IEEE Trans. Syst. Man Cybern. 3(6), 610–621 (1973)
46. Ochs, P., Malik, J., Brox, T.: Segmentation of moving objects by long term video analysis. IEEE Trans. Pattern Anal. Mach. Intell. 36(3), 1187–1200 (2014)

Chapter 3
Scale-Insensitive MSER Features: A Promising Tool for Meaningful Segmentation of Images

Andrzej Śluzek

Abstract Automatic annotation of image contents can be performed more efficiently if it is supported by reliable segmentation algorithms which can extract, as accurately as possible, areas with a certain level of semantic uniformity on top of the default pictorial uniformity of regions extracted by the segmentation methods. Obviously, the results should be insensitive to noise, textures, and other effects typically distorting such uniformities. This chapter discusses a segmentation technique based on SIMSER (*scale-insensitive maximally stable extremal regions*) features, which are a generalization of popular MSER features. Promising conformity (at least in selected applications) of such segmentation results with semantic image interpretation is shown. Additionally, the approach has a relatively low computational complexity ($O(logn \times n)$ or $O(logn \times n \times log(log(n)))$, where n is the image resolution) which makes it prospectively instrumental in real-time applications and/or in low-cost mobile devices. First, the chapter presents fundamentals of SIMSER detector (and the original MSER detector) in gray-level images. Then, relations between semantics-based image annotation and SIMSER features are investigated and illustrated by extensive experiments (including color images, which are the main area of interest).

3.1 Introduction

Image annotation is one of the ultimate tools in large-scale visual information retrieval, e.g. [1, 2]. In general, annotation assigns a number of linguistic captions (tags, labels) to either the whole image or to its selected fragments. The latter category of annotation is considered more informative because geometric distributions of various tags can provide richer descriptions of the image contents (including understanding of activities taking place within the image). Nevertheless, to automatically perform this type of annotation, the images should be segmented into regions

A. Śluzek (✉)
Khalifa University, Abu Dhabi, United Arab Emirates
e-mail: andrzej.sluzek@kustar.ac.ae

© Springer International Publishing AG, part of Springer Nature 2018
H. Kwaśnicka and L. C. Jain (eds.), *Bridging the Semantic Gap in Image and Video Analysis*, Intelligent Systems Reference Library 145,
https://doi.org/10.1007/978-3-319-73891-8_3

which can meaningfully contribute to the annotation decisions. In other words, the regions extracted by the segmentation algorithm should have a decent chance of being semantically distinctive as well.

For popular objects (e.g. human faces and other visual categories often used in various benchmark datasets) numerous dedicated detectors have been developed, so that the corresponding tags can be straightforwardly assigned to the images or to regions segmented by those detectors. Such dedicated detectors are often based on machine learning methods (e.g. [3, 4]).

For unspecified image contents, however, more universal segmentation algorithms should be applied to identify regions which can be prospectively used as (or significantly contribute to) semantically distinctive units.

The range of available segmentation algorithms is very wide (e.g. [5, 6]) with diversified mathematical fundamentals, varying expectations (e.g. full-image segmentation versus background-foreground segmentation where only the foreground regions should be identified) and highly diversified computational complexity (e.g. image thresholding versus active contours or graph-cut methods). Unfortunately, none of the segmentation techniques delivers results which can consistently satisfy requirements of the semantic-based segmentation. In particular, there are no general relations between sophistication/complexity of the segmentation algorithms and the quality (in terms of semantic accuracy) of results.

Figure 3.1 shows a simple example where straightforward thresholding provides results accurately representing the image semantics, while a more sophisticated segmentation algorithm generates rather meaningless regions which require a lot of post-processing before the image can be correctly annotated. Opposite scenarios, where advanced segmentation algorithms produce results better aligned with the semantic-based segmentation are more typical, of course.

Altogether, a number of rather obvious (but often intrinsically contradictive) requirements and recommendations can be listed if we aim to effectively exploit segmentation results in image annotation:

(a)   **(b)**   **(c)**

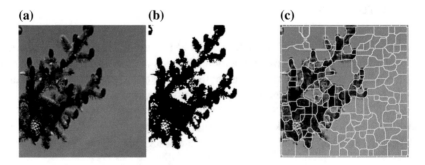

Fig. 3.1 An exemplary image (**a**) in which a simple binary thresholding provides semantically correct segmentation results (**b**), while the watershed segmentation algorithm [8] generates rather meaningless regions (**c**). From [7]

- Segmentation algorithms should be universally applicable to regions of diversified sizes and shapes.
- Segmentation results should be invariant to photometric (e.g. contrast/illumination variations, additive noise, blur, shadows, etc.) and geometric (e.g. scale, perspective projections, occlusions, etc.) distortions.
- Texturization should not affect the segmentation results.
- A clear distinction between foreground and background image components should be provided, if required.
- Segmentation algorithms should be able to exploit both intensity and color data (and additional image dimensions, e.g.depth, if available).
- Low computational complexity of the algorithms, possibly supporting hardware implementation (e.g. specialized *systems-on-chip*).

The most straightforward contradictions exist within the first three points. For example, textured (but otherwise uniform) regions can be confused with collections of evenly distributed small objects, shadows can visually split otherwise uniform regions, etc. More thorough discussions on relations between image segmentation and image semantics can be found in several sources, e.g. in [9].

In this paper, we propose a promising segmentation approach which has its roots in local feature detection. Specifically, the prospective usefulness of *maximally stable extremal regions* (MSER features) and their derivatives in semantic-based segmentation is evaluated.

MSERs were originally proposed for stereo-matching (see [10]) but later found their main applications in content-based visual information retrieval and object tracking (e.g. [11–13]). We argue that although MSERs in their original form are rather poorly suitable for semantic-based segmentation, their generalization, i.e. *scale-insensitive maximally stable extremal regions* (SIMSER features) are a promising tool. Using popular datasets as a benchmark, we show that SIMSER regions are more likely to be semantically distinctive, and they better satisfy the requirements listed above.

As a brief introduction to the fundamentals of this chapter, Sect. 3.2 gives an overview of MSER and SIMSER features, their properties and characteristics. The main analysis and the illustrative experimental examples of SIMSER-based image segmentation are contained in Sect. 3.3, while Sect. 3.4 summarizes the presented results and provides conclusions. Selected computational and algorithmic details are included in the appendix.

3.2 Summary of MSER and SIMSER Features

3.2.1 MSER Features

MSER regions are one of the most popular local features, and are applied in diversified areas of machine vision, mainly in retrieval, detection and tracking tasks. They have

(a)   (b)   (c)

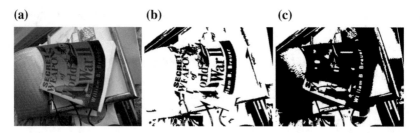

Fig. 3.2 An exemplary image (**a**), and its dark (**b**) and bright (**c**) MSER features. From [7]

been introduced by Matas et al. in [10] but the improved (in terms of computational complexity) variant proposed by Nistér and Stewénius in [14] became more popular.

Informally, *maximally stable extremal regions* are patches within binarized images which are least affected (in terms of their shapes) by gradually changing binarization thresholds.

When a gray-level image is binarized with gradually incrementing threshold values, a family of binary images is formed with correspondingly shrinking white (i.e. above-the-threshold) regions and expanding black (i.e. below-the-threshold) regions. Those regions which are least sensitive to the threshold changes define MSER features. From the practical perspective, MSERs are image fragments which are prominently brighter or darker then their neighborhoods. Figure 3.2 shows an exemplary image and its both bright and dark MSERs. It can be noticed that the shapes of MSER blobs quite accurately correspond to what a human sense of vision would perceive as the most distinctive fragments of the original image.

Formally, MSER features are defined by the local minima of *growth-rate* functions $q(t)$ specified over binary regions created in the process of image thresholding. Given a range of thresholds $t \in T$ and a binary region $Q(t)$ (i.e. a region in the image thresholded at a certain value t) the growth-rate function $q_Q(t)$ is defined by the derivative of the region's area over the threshold level, additionally normalized by the region's area:

$$q_Q(t) = \frac{\frac{d}{dt} \|Q(t)\|}{\|Q(t)\|}, \qquad (3.1)$$

In the actual implementations, the growth-rate functions from Eq. 3.1 defined over continuous threshold values are replaced by one of their discrete approximations:

$$q_Q(t_j) = \frac{\|Q(t_j) - Q(t_{j-1})\|}{\|Q(t_j)\|} \quad \text{or} \quad q_Q(t_j) = \frac{\|Q(t_{j+1}) - Q(t_{j-1})\|}{\|Q(t_j)\|}, \qquad (3.2)$$

where the distance between neighboring thresholds $t_j - t_{j-1}$ defines the threshold increment Δt. For images with 256 levels of intensity, typical values of Δt are within the range $\langle 3; 5 \rangle$, and such values are used within this paper.

Fig. 3.3 System-on-chip detector of MSER features designed according to the architecture reported in [15, 16]

When the threshold level is incremented, the white regions can shrink or disappear, while black regions either expand or merge, so that the families of nested binary regions are always formed. Because of nesting, computational complexity of MSER detection is very low. In the original algorithm proposed in [10] the complexity was $O(n \times log(log(n)))$ (where n is the image resolution) while the algorithm modified in [14] has the linear complexity $O(n)$. Additionally, the regular structure of calculations enables inexpensive hardware (system-on-chip) implementations of MSER detectors with very high throughput (the number of processed images per second) as proposed in [15, 16] (see Fig. 3.3).

MSER features are robust under a wide range of photometric distortions (theoretically, they are invariant to any linear mapping of image intensities) and geometric transformations. In practice, however, they are sensitive to high frequency noise, excessive resolution variations and image texturization. Nevertheless, they are a popular tool, including reported applications in image segmentation (more in Sect. 3.2.3).

In most applications, MSER blobs are approximated by the best-fit ellipses (to calculate keypoint descriptors over such ellipses, e.g. [17]) but this aspect is not further discussed in the paper.

3.2.2 SIMSER Features

The major disadvantages of MSER features are actually related to image rescaling. In particular, blur (which is one of the major effects distorting shapes of extracted MSER blobs) is equivalent to image down-scaling, while noise effects and texture details (which can also affect MSER detection) vary irregularly under image rescaling.

A number of papers have been addressing this issue (e.g. [18–20]) using different approaches, but a recently proposed (see [21]) concept of SIMSER features (*scale-insensitive maximally stable extremal regions*) seems to be the closest follower of the original principles of MSER detection.

In SIMSER detection, instead of detecting regions insensitive only to threshold variations, the algorithm identifies regions which are maximally stable both under the threshold changes and the scale variations (i.e. image blur). It was shown in [21] that such a modification only mildly increases computational complexity of the detection algorithm, while performances are significantly improved and the intuitive notion of *prominent image fragments* is better satisfied by SIMSER blobs than by MSERs.

Formally, SIMSER features are defined by the joint local minima of two *growth-rate* functions. Given an image presented over a range of scales $s \in S$ and binarized using a range of thresholds $t \in T$ (i.e. a family of binary images with diversified binarization threshold and varying levels of smoothness) a region $Q(s, t)$ from any of these binary images can be selected. Then, two growth-rate functions $qt_Q(s, t)$ and $qs_Q(s, t)$ are defined by the partial derivatives of the region's area over t and s dimensions, correspondingly:

$$qt_Q(s, t) = \frac{\frac{\partial}{\partial t} \|Q(s, t)\|}{\|Q(s, t)\|}$$

$$qs_Q(s, t) = \frac{\frac{\partial}{\partial s} \|Q(s, t)\|}{\|Q(s, t)\|} \tag{3.3}$$

A region $Q(s, t)$ is consider a SIMSER feature if both $qs_Q(s, t)$ and $qs_Q(s, t)$ have the local minima there.

The schemes for computing $qt_Q(s, t)$ and $qs_Q(s, t)$ growth-rate functions in the discretized *Threshold* $\times$ *Scale* space are basically similar to Eq. 3.2, i.e.

$$qt_Q(t_j, s_k) = \frac{\|Q(t_j, s_k) - Q(t_{j-1}, s_k)\|}{\|Q(t_j, s_k)\|} \quad \text{or}$$

$$qt_Q(t_j, s) = \frac{\|Q(t_{j+1}, s_k) - Q(t_{j-1}, s_k)\|}{\|Q(t_j, s_k)\|}, \tag{3.4}$$

$$qs_Q(t_j, s_k) = \frac{\|Q(t_j, s_k) - Q(t_j, s_{k-1})\|}{\|Q(t_j, s_k)\|} \quad \text{or}$$

$$qs_Q(t_j, s_k) = \frac{\|Q(t_j, s_{k+1}) - Q(t_j, s_{k-1})\|}{\|Q(t_j, s_k)\|}. \tag{3.5}$$

Computational complexity of SIMSER detection is directly derived from the complexity of MSER detection, which is $O(n)$ in the more efficient variant of the algorithm. Since MSER detection is applied in a range of scales, the numerical complexity is multiplicatively increased by the number of scales.

In practise, the number of scales is proportional to the image resolution. At each subsequent scale, the image is smoothed (e.g. by a Gaussian filter with $\sigma = \sqrt{2}$ which is visually equivalent to halving the image resolution) until the effective resolution falls below the assumed threshold. In [21], the proposed number of scales is

$$1 + \left\lfloor \log_2 (n/64) \right\rfloor , \tag{3.6}$$

where n is the image resolution. For example, for images of VGA resolution 640 × 480 the recommended number of scales is 13.

Therefore, the recommended number of scales can is proportional to $log(n)$ so that the theoretical complexity of the detector would be changed to $O(n \times log(n))$. However, this estimate of complexity does not take into account the detection of local minima of $qs_Q(s, t)$ growth-rate functions (this problem does not exist in MSER detection). Unfortunately, unlike binary regions which are always nested over the changing threshold level, the binary regions over the changing scales do not nest (see the Appendix) so that the minima of $qs_Q(s, t)$ cannot be found straightforwardly. Nevertheless, the computational complexity of $qs_Q(s, t)$ minima detection is also only $qs_Q(s, t)$.

Pseudo-code and more detailed description of SIMSER detection algorithm are included in the Appendix (following specification available in [22]). Therefore, hardware architecture for SIMSER detection can be relatively easily developed from the existing architectures of MSER detector.

Similarly to MSERs, SIMSER blobs can be approximated by the best-fit ellipses.

As reported in [21], SIMSER features inherit many properties of MSERs (including invariance characteristics, the average mnumbers of features in typical images, etc.). However, they have been found superior in many practical aspects. First, SIMSER features are usually better concentrated in the areas of visual prominence. Secondly, they seem to have higher repeatability under image distortions. Finally, they can better identify areas of highly diversified appearances (e.g. textured areas) as long as some visual uniformity exists within such areas. This last property, in particular, makes SIMSERs an attractive tool for image segmentation, as further discussed in the subsequent parts of this chapter.

3.2.3   Segmentation Using MSER Blobs

Both MSER and SIMSER blobs are actually image regions with some level of visual uniformity, which naturally links them with image segmentation. Therefore, some works report applications of MSER features in segmentation problems. For example, a partially supervised video segmentation is used in [23]. The user indicates in the first frame the approximate location of interest, and the algorithm identifies MSER regions around that location both in the first frame and in the subsequent ones.

In [24], MSER blobs are clustered using a weighted graph of MSER nodes, with a matrix of similarity defined by color or intensity closeness between pixels of adjacent blobs. Then, graph partitioning is used to divide the image into the required number of segments.

The method proposed in [25] makes use of a multi-scale structure of image color regions, where core regions (MSER blobs) gradually absorb the most similar non-core areas until the whole image is segmented. In [26], MSER-like structures are applied as a supplementary tool.

Altogether, it can be concluded from the published reports that MSER detection provides a kind of useful image segmentation, but the semantic significance of those segmentation results is often very limited, and complex post-processing steps are needed to obtain more meaningful image partitioning from MSER blobs.

Examples in Fig. 3.4 clearly illustrate limitations of MSER-based segmentation. Note that black areas (i.e. not included into MSER blobs) are considered non-segmentable background.

In the first case (the image is from a dataset of manually semantically segmented images, [27]) the results superficially look acceptable, but it can be noticed that some of semantically uniform areas actually consist of large numbers of MSER blobs. Altogether, especially because dark and bright MSER blobs overlap (which happens pretty frequently, and may slightly distort the visual perception of segmentation results) the overall impression is that generally no semantic interpretation can be assigned to individual blobs. It can be also noticed that sometimes MSER blobs of the same category (i.e. either dark or bright) are nested, but this effect is more plausible and does not create any semantic confusion.

In case of the second image (from a popular dataset[1]) the segmentation results obviously have no semantic meaning.

In the next Sect. 3.3, we show that much more satisfactory results can be obtained using SIMSER blobs as the basic segmentation units.

3.3   SIMSER-Based Image Segmentation

The hypothesis to be investigated in this section is whether the direct (i.e. without any further post-processing and/or blob mergers) outputs of SIMSER detection can provide semantically meaningful image segmentation. The underlying assumption is, therefore, that the image areas outside SIMSER blobs are considered non-segmentable background, i.e. the whole procedure falls into the category of *foreground-background* segmentation.

As an introduction, the SIMSER-based equivalents of Fig. 3.4 results are shown in Fig. 3.5. In the first case, almost all SIMSER blobs (especially if the nested blobs are taken as a single unit) can be tagged as 'corals', 'coral reef' or 'deep water'. Even the difficult second image has now two regions which can be labeled as 'vegetation' and 'darker part of the wall'.

---

[1]http://www.robots.ox.ac.uk/~vgg/research/affine/.

Fig. 3.4   Two exemplary images and their MSER-based segmentation

Fig. 3.5   SIMSER-based segmentation of images from Fig. 3.4

Although SIMSER features have been introduced and preliminarily evaluated in [21], the analysis presented there was focusing on visual information retrieval tasks so that further analysis is needed to better understand performances of SIMSER features in segmentation problems. The following sub-sections present some of our findings.

3.3.1 Image Smoothing in SIMSER Detection

The standard scheme of image smoothing in the discrete space of scales is to use Gaussian filters with $\sigma = \sqrt{2}$ (which is equivalent to halving the image resolution, e.g. [28]). In such scenarios (and with the number of scales defined by Eq. 3.6) the numbers of SIMSER features detected in typical images were reported almost the same as the numbers of MSERs.

We re-examined this estimate on much larger collections of images (including datasets exploited in this paper) and have found that the numbers of SIMSERs are actually significantly lower than the numbers of MSERs in the images. The average number of SIMSERs is typically only 50% of the number of MSERs in the same image. To some extent, these differences can be attributed to a slightly different numerical scheme used for MSER and SIMSER detection in this paper (see Eqs. 3.2, 3.4 and 3.5) but, nevertheless, the numbers of SIMSERs are still too high, if we expect that most of them carry some semantics.

Therefore, without changing the number of scales, we have applied image *over-smoothing* in the scale space, i.e. Gaussian filters with larger values σ are used at transitions between neighboring scales. With the increased σ, the numbers of detected SIMSERs are gradually decreasing (we do not have a convincing theoretical explanation of this phenomenon yet). Based on extensive tests, we eventually selected $\sigma = 3$, for which the average number of SIMSER blobs is only 9.3% is MSER blobs (the results presented in Fig. 3.5 are obtained using this value of σ). The corresponding statistics are provided in Table 3.1.

To further illustrate the effects of over-smoothing, three examples are given in Fig. 3.6. Many SIMSERs detected with the smaller σ do not carry any clearly identifiable semantics (although some of them provide semantics not available at the larger σ, e.g. the bird eye in the bottom row).

Table 3.1 Average numbers of feature blobs in images (based on over 1000 outdoor images)

Feature	Average number
MSER	757.2
SIMSER ($\sigma = \sqrt{2}$)	281.2
SIMSER ($\sigma = 3$)	70.4

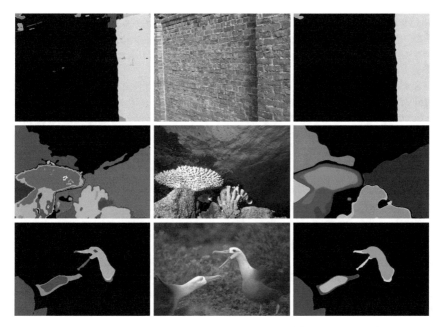

Fig. 3.6 Examples of SIMSER-based segmentation using $\sigma = \sqrt{2}$ (left) and $\sigma = 3$ (right)

3.3.2 SIMSER Detection in Color Images

Currently, almost all multimedia collections and databases (e.g. internet resources) contain mostly color images (with rather few exceptions reserved for historical records and images acquired using non-standard capturing devices) or even images of higher dimensionality (e.g. RGB-D images). In such images, segmentation techniques can exploit both unique characteristics of individual channels and combined properties of the channels. The same applies to SIMSER-based segmentation.

Even though all images of our experiments are RGB, the results presented in the previous examples are based only on gray-level images, i.e. on combined properties of three channels. Nevertheless, SIMSER detection and SIMSER-based segmentation can be easily performed in each individual channel so that certain semantics hidden in individual colors can be better represented by the segmented blobs.

To illustrate potential benefits of individual channels in SIMSER-based segmentation, a number of examples are shown in Fig. 3.11 (at the end of paper). In some cases, individual channels provide very similar results to gray-level images, sometimes fewer SIMSER blobs are extracted from individual channels, but there are also cases where certain semantically distinctive blobs are extracted from a particular channel only.

Three interesting cases can be pointed out. In Fig. 3.11g, a semantically unique region (which can be spotted in the original image and tagged as lighter part of the wall) is extracted as a SIMSER blob in the *red* channel only. A stained (wet?) top part of the wall is found only in the *blue* channel (Fig. 3.11m). Also, the *red* channel is the only one in which the whole water area is identified as non-segmentable background (Fig. 3.11h).

Such results should not be surprising, because many segmentation techniques exploit multi-spectral properties of processed images. Nevertheless, it is comforting to acknowledge that an extremely simple algorithm based on rather different principles (MSER/SIMSER detectors have their roots in images matching and retrieval) can provide results with similar characteristics.

3.4   Concluding Remarks

The chapter overviews a low-complexity image segmentation methodology which has a potential of narrowing the gap between pictorial-only segmentation and segmentation with some semantics embedded.

The method is developed from region-based local features used for image matching and retrieval. Starting from MSER features (which have been found rather useless in semantic image segmentation, unless supported by complicated post-processing operations) we eventually focus on *scale-insensitive maximally stable extremal regions* (SIMSER features).

In previous works, SIMSER features have been reported superior to MSERs in image matching and retrieval tasks, [21]. In this chapter, we argue that SIMSERs can also much better than MSERs contribute to semantics-oriented segmentation and, therefore, to automatic image annotation. In particular, regions extracted by the SIMSER detector can correspond to both large objects with complicated textures (as long as these objects are visually uniform in some sense) and very small objects visually protruding from the background. This is possible because SIMSER blobs are insensitive both to image thresholding and image rescaling. Thus, large textured objects can be identified at larger scales (at which the texturization details are smoothed) while small objects (even individual components of the textures, if prominent enough) would be detected at finer scales.

The proposed method belongs to the foregroud-background segmentation techniques, where SIMSER blobs are considered foreground objects with some semantics associated. Usually, the numbers of semantically distinctive components within a single image are rather limited. Therefore, the number of SIMSER features should not be excessively large (i.e. they should be much smaller than, for example, typical numbers of MSER features in the same images). This is achieved, without changing the computational structure of SIMSER detector, by using *over-smoothing*, i.e. by widening the Gaussian filter used for smoothing the images when the image scales are gradually changed. This results in a ten-fold reduction in typical numbers of extracted blobs so that most of them can have a meaningful semantic interpretation.

In case of color images, the method can be applied both to individual channels and to gray-level copies so that the chances of revealing semantics hidden behind more complicated visual appearances are enhanced.

Computational costs of the method are very low and its algorithmic structure is very regular. Actually, the SIMSER detection algorithm can be rather easily converted into SoC (*system-on-chip*) implementation by modifying the existing chip for MSER detection.

In general, performances of the SIMSER-based segmentation in the context of semantic accuracy and correctness can be evaluated primarily qualitatively. Nevertheless, certain quantitative results are available as well. In [29], the preliminary version of the method was evaluated on a dataset of semantically segmented images [27], using the following measure of dissimilarity between overlapping regions:

$$dis(A, B) = 0.5 \left(\frac{\|A - A \cap B\|}{\|A\|} + \frac{\|B - A \cap B\|}{\|B\|} \right), \tag{3.7}$$

where A and B are the compared region (note that 0 value indicates identical regions while 1 means disjoined regions).

The manually outlined semantic regions were compared, using Eq. 3.7, to SIMSER blobs maximally overlapping with those region. Over the whole dataset, the average values of so defined dissimilarity between the ground-truth semantic regions and their best SIMSER approximations is 0.295. To provide the intuitive meaning of this number, Fig. 3.7 gives examples of overlapping rectangles with such dissimilarity values.

These results are not conclusive yet, neither they indicate perfect correspondences between semantic-based segmentation results and SIMSER detection, but they preliminarily suggest that semantically uniform visual regions can often be quite accurately approximated by SIMSER blobs.

The continuation of the presented works will focus mainly on further experimental verification of the SIMSER-based segmentation performances in diversified scenarios (including real-time applications, for which the method seems particularly suitable because of its extremely low computational complexity) and on diversified classes of images. Development of hardware solutions for SIMSER detection is also considered, subject to availability of adequate resources.

Fig. 3.7 Examples of overlapping rectangles with the dissimilarity value 0.295

Acknowledgements Some results presented in this paper have been supported by the ATIC-SRC Center within Energy Efficient Electronic Systems contract 2013-HJ-2440 for the task A Low-Power System-on-Chip Detector and Descriptor of Visual Keypoints for Video Surveillance Applications.

Appendix

The Appendix contains details of computational steps in SIMSER detection, focusing on the prospective hardware or hardware-supported implementations. However, such details cannot be fully explained without an insight into the detection of MSER features. Thus, the included information (a summary of the results presented in [22]) covers most important facts on architectures used in MSER and SIMSER detection, as well as architectures specifically proposed for SIMSER detection only.

Detection of local minima in the threshold space

At each threshold level, the binary image of $M \times N$ size is represented by three data structures:

- *Seed matrix* of regions SM (of the same size as the image) with the initial content $SM_{i,j} = M \times (i - 1) + j$, i.e. each pixel is a seed for itself. After processing, $SM_{i,j} = K$, where K indicates the initial pixel (seed) of the region to which (i, j) pixel belongs.
- *Region Size* matrix RS (of the same size) specifying the size of region to which each (i, j) pixel belongs. Initially, $RS_{i,j} = 1$, i.e. each pixel is a separate region of unit size.
- *Map-of-regions* array, which for each image region lists its seed, the binary color and the size.

A small binary image and the final contents of its SM and RS matrices are shown in Fig. 3.8, while its *Map-of-regions* is given in Table 3.2.

Given such representations for the sequence of binary regions over three neighboring threshold levels (note that such regions are always nested) the local minima of q_Q (see Eq. 3.2) and qt_Q (see Eq. 3.4) growth-rate functions can be straightforwardly identified. In other words, MSER regions can be detected or SIMSER candidates (i.e. the regions which satisfy the local minimum criterion in the threshold space) can be pre-selected.

Fig. 3.8 A small binary region and its final SM and RS matrices

$$\begin{bmatrix} 1 & 2 & 2 & 2 & 2 \\ 1 & 2 & 1 & 2 & 10 \\ 1 & 1 & 1 & 2 & 10 \\ 2 & 2 & 2 & 2 & 10 \\ 2 & 2 & 10 & 10 & 10 \end{bmatrix}$$

Final SM

$$\begin{bmatrix} 6 & 13 & 13 & 13 & 13 \\ 6 & 13 & 6 & 13 & 6 \\ 6 & 6 & 6 & 13 & 6 \\ 13 & 13 & 13 & 13 & 6 \\ 13 & 13 & 6 & 6 & 6 \end{bmatrix}$$

Final RS

Table 3.2 *Map-of-regions* for Fig. 3.8 image

	Region 1	Region 2	Region 3
Seed	1	2	10
Size	6	13	6
Color	Dark	Bright	Dark

Fig. 3.9 An example of not nested (overlapping) black and white regions over two neighboring scales (smoothing removes sharp extremes, both dark and bright)

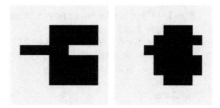

Detection of local minima in the scale space

To identify SIMSER blobs, the regions pre-selected as the local minima in the threshold space should also be confirmed as the local minima in the scale space, i.e. the minima the second growth-rate function qs_Q (see Eq. 3.5). To verify this, two operations are needed:

- The original input image should be repetitively processed by a smoothing filter. This is just a convolution with the filter kernel, i.e. the operation which can be straightforwardly into hardware. Its computational complexity is $O(n)$.
- The correspondences between binary regions in the neighboring scales should be established and, based on that, the values of qs_Q growth-rate evaluated. This is not a straightforward operation because binary regions over a sequence of scales often do not nest (a simple example is shown in Fig. 3.9).

To solve this problem, the following pseudocode is proposed (its less effective variant which, nevertheless, clearly indicates $O(n)$ complexity of the algorithm was given in [21]):

EVALUATION OF qs_Q GROWTH- RATE FUNCTION

```
Input Im1(M,N), Im2(M,N)
% two binary-images matrices at two neighbouring scales (M*N size)
Input RS1(M,N), RS2(M,N)
% two region-size matrices (M*N size)
Input MoR1(K1), MoR2(K2)
% two maps-of-region
Storage MX(K1,K2)<- zeros
% size of region intersections matrix
Storage next(K1)<- zeros
% reion correspondences
Storage gs(K1)<- zeros
```

```
% growth-rate values

for i = 1:M
 for j = 1:N
  if Im1(i,j)==Im2(i,j)
   MX(RS1(i,j),RS2(i,j))++
  endif
 endfor
endfor
for k = 1:K1 min = LARGE VALUE; indx = 0;
 for l = 1:K2
  temp = MoR1(k).size + MoR2(l).size 2*MX(k,l)
  if temp < min && MX(k,l) > 0
   min = temp; indx = l;
  endif
 endfor
 gs(k) = min/MoR1(k).size; next(k) = indx;
endfor
```

The scheme takes two binary images (at the same threshold but at the neighboring scales) their *RS* and *SR* matrices, and their *maps-of-regions* (see above). For each binary region at the current scale, the identifier of the next-scale region is found, and the value of the growth-rate function qs_Q is evaluated. Therefore, the changes of qs_Q can be tracked over the scales, and the local minima can be easily found.

In this way, all operations needed to identify SIMSER features are completed.

As an example, a pair of binary images from two neighboring scales is shown in Fig. 3.10, and the corresponding results of the above operations are included in Table 3.3. In this example, *Region 4* has the best chance to be a local minimum (with the smallest value of qs_Q). To confirm that, however, similarly computed values of qs_Q for *Region C* (which is the correspondence of *Region 4* in the next scale) and for the corresponding region in the previous scale, should be larger (Fig. 3.11).

Altogether, it can be concluded that SIMSER detection architecture is a relatively simple extpansion of the MSER detection architecture, so that hardware implementation of SIMSER detector is a feasible task.

Fig. 3.10 Computing qs_Q growth-rate function in the scale space. The left image is in the current scale, while the right one in the next scale

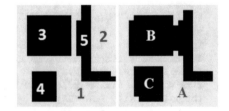

Table 3.3 qs_Q processing for Fig. 3.10 images. MX values are the region intersection sizes

Regions	A(white, size = 241)	B(black, size = 118)	C(white, size = 41)	qs_Q, next reg.
1(white, size = 186)	$MX = 169$	$MX = 0$	$MX = 0$	0.53, A
2(white, size = 66)	$MX = 66$	$MX = 0$	$MX = 0$	2.65, A
3(black, size = 72)	$MX = 0$	$MX = 71$	$MX = 0$	0.68, B
4(black, size = 30)	$MX = 0$	$MX = 0$	$MX = 29$	0.43, C
5(black, size = 46)	$MX = 0$	$MX = 43$	$MX = 0$	1.69, B

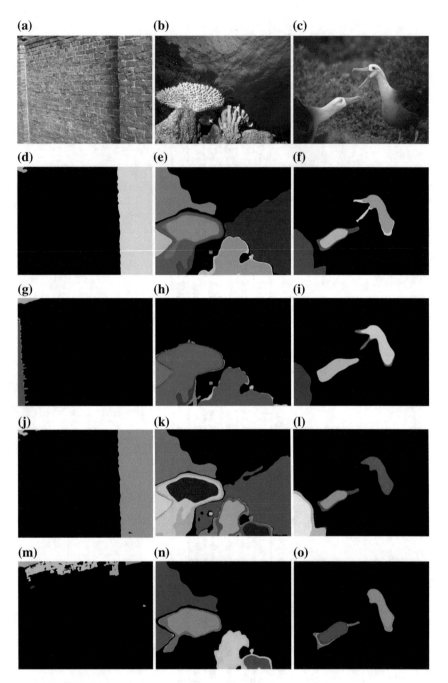

Fig. 3.11 Original images (**a,b,c**) and the SIMSER-based segmentation results obtained from: *grey-level* copies (**d,e,f**), *red* (**g,h,i**), *green* (**j,k,l**) and *blue* (**m,n,o**) channels

References

1. Hadbury, A.: A survey of methods for image annotation. J. Vis. Lang. Comput. **19**, 617–627 (2008)
2. Liu, Y., Zhang, D., Lua, G., Ma, W.-Y.: A survey of content-based image retrieval with high-level semantics. Pattern Recogn. **40**, 262–282 (2007)
3. Viola, P., Jones, M.J.: Robust real-time face detection. Int. J. Comput. Vis. **57**, 137–154 (2004)
4. Zhou, B., Lapedriza, A., Xiao, J., Torralba, A., Oliva, A.: Learning deep features for scene recognition using places database. In: Ghahramani, Z., et al. (eds.) Advances in Neural Information Processing Systems 27 (NIPS 2014), pp. 487–495. Curran Associates, Inc. (2014)
5. Pal, N.R., Pal, S.K.: A review on image segmentation techniques. Pattern Recogn. **26**, 1277–1294 (1993)
6. Zaitouna, N.M., Aqel, M.J.: Survey on image segmentation techniques. Procedia Comput. Sci. **65**, 797–806 (2015)
7. Śluzek, A.: Local Detection and Identification of Visual Data: Selected Techniques and Applications. LAP, Saarbrucken (2013)
8. Belaid, L.J., Mourou, W.: Image segmentation: a watershed transformation algorithm. Image Anal. Stereol. **28**(2), 93–102 (2009)
9. Thoma, M.: A survey of semantic segmentation. https://arxiv.org/pdf/1602.06541. Accessed 27 Apr 2017
10. Matas, J., Chum, O., Urban, M., Pajdla, T.: Robust wide baseline stereo from maximally stable extremal regions. In: Proceedings of British Machine Vision Conference BMVC 2002, pp. 384–393 (2002)
11. Wu, Zh., Ke, Q., Isard, M., Sun, J.: Bundling features for large scale partial-duplicate web image search. In: Proceedings of 2009 IEEE Conference Computer Vision & Pattern Recognition CVPR 2009, vol. 1, pp. 25–32 (2009)
12. Donoser, M., Bischof, H.: Efficient maximally stable extremal region (MSER) tracking. In: Proceedings of 2006 IEEE Conference Computer Vision & Pattern Recognition CVPR 2006, vol. 1, pp. 553–560 (2006)
13. Gómez, L., Karatzas, D.: MSER-based real-time text detection and tracking. In: Proceedings of 22nd International Conference on Pattern Recognition ICPR 2014, pp. 3110–3115 (2014)
14. Nistér, D., Stewénius, H.: Linear time maximally stable extremal regions. In: Proceedings of 10th European Conference ECCV 2008. vol. 2, pp. 183–196 (2008)
15. Salahat, E., Saleh, H., Sluzek, A., Al-Qutayri, M., Mohammad, B., Elnaggar, M.: Architecture and method for real-time parallel detection and extraction of maximally stable extremal regions (MSERs). US Patent 9,311,555, 12 Apr 2016
16. Salahat, E., Saleh, H., Sluzek, A., Al-Qutayri, M., Mohammad, B., Elnaggar, M.: Hardware architecture for real-time extraction of maximally stable extremal regions (MSERs). US Patent 9,489,578, 8 Nov 2016
17. Mikolajczyk, K., Schmid, C.: A performance evaluation of local descriptors. IEEE Trans. PAMI. **27**, 1615–1630 (2005)
18. Forssén, P.-E., Lowe, D.G.: Shape descriptors for maximally stable extremal regions. In: Proceedings of 11th IEEE International Conference on Computer Vision ICCV 2007, pp. 1–8 (2007)
19. Kimmel, R., Zhang, C., Bronstein, A.M., Bronstein, M.M.: Are MSER features really interesting? IEEE Trans. PAMI. **33**, 2316–2320 (2011)
20. Martins, P., Carvalho, P., Gatta, C.: On the completeness of feature-driven maximally stable extremal regions. Pattern Recogn. Lett. **74**, 9–16 (2016)
21. Śluzek, A.: Improving performances of MSER features in matching and retrieval tasks. In: Proceedings of 14th European Conference ECCV 2016. vol. LNCS 9915, pp. 759–770 (2016)
22. Śluzek, A., Saleh, H.: Algorithmic foundations for hardware implementation of scale-insensitive MSER Features. In: Proceedings of 59th International Midwest Symposium Circuits & Systems MWSCAS 2016, pp. 1–4 (2016)

23. Donoser, M., Bischof, H., Wiltsche, M.: Color blob segmentation by MSER analysis. In: Proceedings of IEEE International Conference on Image Processing ICIP 2006, pp. 757–760 (2006)
24. Gui, Y., Zhang, X., Shang, Y.: SAR image segmentation using MSER and improved spectral clustering. EURASIP J. Adv. Sig. Process. **83** (2012)
25. Oh, I.S., Lee, J., Majumder, A.: Multi-scale image segmentation using MSER. In: Proceedings of 15th International Conference CAIP 2013, vol. II, pp. 201–208 (2013)
26. Wang, G., Gao, K., Zhang, Y., Li, J.: Efficient perceptual region detector based on object boundary. In: Proceedings of 22nd International Conference on Multimedia Modeling MMM 2016, vol. II, pp. 68–78 (2016)
27. Li, H., Cai, J., Nguyen, T.N.A., Zheng, J.: A benchmark for semantic image segmentation. In: Proceedings of IEEE International Conference Multimedia and Expo ICME 2013 (2013)
28. Lindeberg, T.: Feature detection with automatic scale selection. Int. J. Comput. Vis. **30**, 77–116 (1998)
29. Śluzek, A.: MSER and SIMSER regions: A link between local features and image segmentation. In: Proceedings of International Conference on Computer Graphics & Digital Image Processing CGDIP 2017, Article 15 (2017)

Chapter 4
Active Partitions in Localization of Semantically Important Image Structures

Arkadiusz Tomczyk

Abstract In this chapter active partitions, a generalization of active contours concept to other than pixel-based image representations, is presented. Active contours are methods where optimal, with respect to a given objective function, contours are sought in the images. Their main advantage is fact that they are able to use any additional expert knowledge while analyzing the images. It is of special importance if in the image itself there is no sufficient visual information allowing for proper interpretation of its content. That knowledge can be incorporated into the search process by proper selection of contour model, soft constraints in energy function or hard constraints in an optimization procedure. All those advantages are preserved in active partitions where image content is described not with pixels but with other set of semantically more informative elements. Consequently, in active partitions not an optimal contour is sought but optimal partition of given element set is looked for. The change of image content description is advantageous as well. It reduces the size of search space and allows humans to express their knowledge in more intuitive way.

4.1 Introduction

There are many different image segmentation techniques that can be directly or indirectly applied to the tasks of object localization within an image. The main limitation of the classic methods, such as thresholding or region growing, is that they consider only what is available in the image itself, failing to utilize external knowledge about the structure of interest. Such knowledge is crucial in those tasks where the image itself contains insufficient information for proper semantic interpretation of its content. A typical example here is radiological image interpretation, which requires adequate anatomical knowledge, without which it would be impossible to

A. Tomczyk (✉)
Institite of Information Technology, Lodz University of Technology,
ul. Wolczanska 215, 90-142 Lodz, Poland
e-mail: arkadiusz.tomczyk@p.lodz.pl

© Springer International Publishing AG, part of Springer Nature 2018
H. Kwaśnicka and L. C. Jain (eds.), *Bridging the Semantic Gap in Image
and Video Analysis*, Intelligent Systems Reference Library 145,
https://doi.org/10.1007/978-3-319-73891-8_4

distinguish between organs that have a similar representation of tissues in an image modality under consideration.

A possible solution to that problem is provided by the active contour techniques. This group of methods operate under the assumption that the space of contours, unambiguously identifying objects in the image, is defined. The main objective is to find an optimal contour within that space by proper selection of the objective function (energy) and the optimization algorithm (evolution). The active contour model owes its name to the fact that optimization is usually an iterative procedure, which results in a change of the contours shape after each iteration.

The external knowledge can be incorporated into the active contour procedure in several ways:

- Proper selection of the contour model can eliminate semantically incorrect solutions.
- Proper constraints imposed on the optimization process can prevent obtaining unacceptable solutions.
- Proper components of the energy function can penalize solutions that do not reflect our expectations.

Active partitions can be considered as a generalization of active contours. During the localization process the contours divide the set of image pixels into two subsets representing the object and the background. Such a partition can be defined, however, for any set of elements describing the image content, e.g. superpixels, line segments, ellipses, etc. The change of image description can significantly reduce the space of analyzed primitives without losing important semantic information, which can be still encoded in the attributes assigned to those primitives. The chief advantages of this approach are as follows:

- The reduced image representation enables the construction of solutions that resemble more closely a conscious image analysis process specific to human beings.
- Incorporation of external knowledge seems to be more natural.
- Reduced search space allows the use of more computationally demanding optimization algorithms.
- More sophisticated optimization algorithms provide the ability to avoid problems with proper selection of initial solutions.

In this chapter, the concept of active partitions is presented and illustrated. Issues regarding medical image analysis fall outside the scope of this study. For simplicity reasons, the author focuses exclusively on the problem of warning road sign localization. The chapter is organized as follows. In Sect. 4.2 a short overview of the active contour techniques is provided, together with the methods of external knowledge incorporation. Section 4.3 introduces the basic concepts of active partitions. Section 4.4 presents a simplified example of active partition application. The chapter concludes with a short summary in Sect. 4.5.

4.2 Background

As already mentioned, the process of object localization in active contours is formulated as an optimization problem, where the objective function (energy) expresses the expectations about the structure of interest. Thus, the energy, if properly defined, should assign to the contour its optimal value (usually minimal) only when the contour represents the object of interest. It is obvious that this evaluation should consist of at least two types of components:

- external components taking into account the position of the contour in the image (they can evaluate whether the contour lies on the visible boundaries or circumscribes the region with desired characteristics, etc.)
- internal components taking into account the characteristics of the contour itself (they can evaluate local contour smoothness, global contour shape, etc.).

In the literature, there are many variants of active contours, each adopting a different contour model, which in turn determines the formulation of energy function components and imposes specific contour evolution strategies. In this section, a short review of active contour techniques is presented, which is followed by a discussion of the methods for encoding knowledge about the expected contour characteristic.

4.2.1 Active Contours

The term active contour was first proposed in [1] by Kass, Witkin and Terzopoulos, who described the *snakes* model, in which the contour was represented by a parametric curve in the image plane. Since contour parametrization is a function, the energy is a functional, and to analytically find an optimal contour, the calculus of variations needs to be used. The application of Euler-Lagrange equations leads in this case to the system of partial derivative equations. Its numerical solution, which requires contour discretization (the contour is transformed into a polygon), results in an iterative process of optimal solution finding. Since the position of contour points is modified at each iteration, the whole process can be interpreted as the movement of the contour under the influence of some internal and external forces. This provides the ability to avoid an explicit definition of the energy function and replace it by a direct definition of the forces modifying the contour according to user expectations.

Another popular variant of active contours is the *geometric active contours* approach. It was proposed simultaneously by Malladi, Sethian and Vemuri in [2] and by Casseles, Catte and Dibos in [3]. In those methods, the internal parametrization of the curves is not considered since it does not influence the contours shape. Consequently, only forces normal to the contour are taken into account. At first, the energy function was not expressed explicitly. It was added in *geodesic active contours* by Casseles, Kimmel and Sapiro in [4] and by Yezzi, Kichenassamy, Kumar, Olver and Tannenbaum in [5]. Thus, contour evolution is usually defined directly by the forces. Using a level-set approach ([6]), where the contour is defined by sections of 3D surface, it is possible to obtain contours of different topology (describing separate regions or those with holes).

Another interesting solution is offered by *active shape models*, first described by Cootes and Taylor in [7]. Here, the contour is represented by a set of characteristic points, which need not compose a polygon. The possible relative positions of those points are statistically trained before evolution (the so called *point distribution model*) on the basis of images that have previously been manually marked. The evolution itself comprises two operations, performed at every iteration. First, locally optimal positions of the characteristic points are sought. This operation usually takes into account the expected image profile around this point. Next, the final position of the points is estimated, allowing only three geometrical modifications of the whole shape (translation, rotation and scale) and some local shape modifications that do not violate *point distribution model* constraints.

A completely different assumption was made by Grzeszczuk and Levin in [8]. Their method, *Brownian strings*, represents the contour in a linguistic way using a chain of directions describing how to move along the contour (the contour lies in the cracks between pixels). Another non-standard element introduced by this method is the optimization technique that it employs to detect the optimal contour. In this case, the *simulated annealing* algorithm is used to avoid problems with precise localization of the initial contour. Its application requires, however, a suitable choice of local contour modifications performed at every iteration. Due to the specificity of the contour model, those modifications are quite complex. The same optimization technique, but another contour model, was applied by Tomczyk and Szczepaniak in the *potential active contours*, proposed in [9]. In this method, the contour is defined by a set of potential field sources. There are two types of those sources and the contour lies where the summary potentials of both types are equal. The evolution of the contour requires potential source modifications which in this case involve changes in their location as well as in the parameters that control the generated potential field characteristic. The optimization technique applied in both those methods provides significant flexibility in defining energy functions, since there are no special requirements as to their form (they need not be differentiable).

In literature, many other variants of active contours can be found. A comparative study can be found in [10, 11]. The choice of specific variant depends on considered application. If the topology of the sought region may change, *geometric active contours* should be used. There are modifications of *snakes* that allow to change region topology but their implementation is less elegant. *Snakes* are good option if contour can be initialized relatively close to the optimum of the energy functions. Otherwise, methods allowing to explore the whole search space, like *Brownian strings* or *potential active contours*, should be considered. The former gives full flexibility of shape description, whereas the latter will be useful if smooth, rounded shapes are to be found. If the sought shapes do not differ too much (mainly in their position, orientation or scale), then properly trained *active shape models* will be the best choice.

For reasons of space, this section focuses further only on those active contour variants that use specific external knowledge regarding the objects of interest to enhance detection. These approaches are described in more detail below.

4.2.2 Knowledge

The ability to incorporate the external knowledge into the process of object localization is a fundamental advantage of active contour techniques. There are three possible elements where the expectations about the structure of interest can be expressed:

- Contour model—Knowledge encoded in the contour model makes it possible to reduce the search space if specific properties of the object are known, for example in *potential active contours* where the space of describable contours contains only smooth and rounded shapes. Additionally, the achievable degree of roundness can be controlled by a number of potential sources. This can be observed also in the models that use *Fourier descriptors* presented in [12, 13] and *splines* discussed in [14, 15]. Smoothness, however, is not the only requirement that can be encoded in the model. In *active rays* ([16]) it assumed that not all the concave shapes need to be described. In this case, a distance to a fixed point in the image plain enables the description of all the desired shapes.
- Evolution strategy—If the energy function is explicitly given and some general purpose optimization technique is used, then knowledge can be used to add hard constraints forbidding certain contour modifications. Typical examples include the *point distribution model* used in *active shape models* and some specific solution generators used in methods that apply the *simulated annealing* algorithm. If the energy need not be specified explicitly and the evolution strategy is designed directly, then knowledge is encoded during the design process. A good example are forces and force fields defined in *snakes* or *geometric active contours*. In [1], for instance, volcano and spring forces were described, whereas in [17] a template force was added to keep a desired shape of the contour.
- Energy function—The expectations about the contour are typically expressed as soft constraints. They can be encoded in both internal and external components, and may have either local or global character. A standard internal local expectation is contour smoothness. In *snakes* it is expressed by the characteristic of curve parametrization derivatives. Another approach imposes the reduction of contour length. External local expectations focus mainly on the image characteristics on both sides of the contour. Global expectations concern usually the contours shape and the characteristics of the region inside and outside of the contour. The latter can be found in *active regions* ([18]) and *active appearance models* ([19]).

A separate problem is the acquisition and representation of the required knowledge. It is not a trivial task but its detailed discussion falls outside of the scope of this chapter. Let us only mention that many approaches use a kind of training procedures where the selection of optimal active contour parameters relies on manual pre-localization of objects. Such procedures can be found in *active shape models* or *active appearance models*, in *Brownian strings* or in *potential active contours*.

4.3 Active Partitions

Although the concept of contour is intuitively clear, and has been the subject of many practical implementations within the framework of the active contour model, there is no single, universally applicable formal definition of this term. An attempt to formally describe the concept of contour was made, among others, in [20]. The basic feature of the contour is its ability to unambiguously indicate which part of the image reflects the object described by the contour and which part constitutes the background. In other words, the contour possesses the ability of dividing the image pixel set into two partitions.

In practical applications, however, operating on contours in a pixel space is problematic. The main issue is the cardinality of the pixel set, since the number of pixel subsets (possible partitions) grows exponentially with the increasing size of the image. Active contours try to tackle this problem in different ways, as described in the previous section. A proper definition of the contour model and evolution constraints can reduce the space of available partitions. Moreover, appropriate contour initialization, such that makes it relatively close to the optimal solution, can allow one to use simpler optimization techniques, guaranteeing that the desired structure is detected. Another issue connected with pixel representations is the difficulty of defining contour energy, as it often requires defining energy components at a pixel level as well. This is something of a pitfall, which also manifests itself further while defining evolution strategies, when potential and force fields are required, for example, in *snakes* and *geometric active contours*. In such a situation, the process of higher-level (global) knowledge incorporation in the localization procedure becomes significantly more difficult.

Because of those reasons, active partitions, a generalization of active contours, was proposed in [21–23]. In that approach, the image is not represented by a set of simple pixels but by a reduced set of more complex, spatially localized elements $E = \{e_1, \ldots, e_N\}$. Naturally, the term contour, understood as a line that separates the object elements from the background elements, is hardly applicable in this context. That is why in active partitions instead of the optimal contour, the optimal partition $P = \{E_O, E_B\}$ is sought directly where $E_O \subseteq E$ and $E_B \subseteq E$ represent object and background elements, respectively, under the assumption that $E_O \cup E_B = E$ and $E_O \cap E_B = \emptyset$.

4.3.1   *Representation*

Although active partitions do not assume anything about the nature of the elements
E, to make the method more natural, it is good to refer to observations of the human
visual system ([24]). From that perspective, it seems to be obvious that humans do
not analyze images directly at a pixel level, focusing rather on local similarities
(homogenous regions) and discontinuities (region borders). To reflect this observation, superpixels and line segments were proposed to represent image content in [21,
22], respectively. Examples of such representations of the images considered in this
chapter are presented in Fig. 4.1.

In [21], to generate a superpixel representation, the *simple linear iterative clus-*
tering SLIC algorithm was used ([25]). It is an adaptation of the k-means clustering
algorithm with a properly defined pixel metric. This representation was used to find
regions representing the interior of the left and right heart ventricle in CT images. To
avoid problems with insufficient image information (heart muscle grows into heart
chambers) the requirement for the partition of a minimum border size was added.
This approach was adapted from *snakes* method. The *simulated annealing* method
was used as an optimization algorithm, with a solution generator ensuring that only
connected partitions were generated. In [22], the content of mammograms and road
scenes was described using a modified *line segment detector* LSD algorithm ([26]).
The line segments reflected the areas of the image where a significant difference of
pixel intensity on both sides of those segments was observed. This representation
was used to localize circular and triangular regions, some of which might indicate
possible circumscribed lesions or warning signs, respectively. In this case, a heuristic
search was proposed to reduce the space of the analyzed subsets of segments. The
energy function was employed to evaluate the matching degree between a current
solution and a given template.

An alternative region-based representation was presented in [23]. In that work, the
MR knee images were represented by ellipses describing the regions of a similar color.
Ellipses were generated using the *cross-entropy clustering* CEC algorithm ([27]).
This helped to reduce the number of considered elements required for the correct
localization of elongated structures forming the fragments of articulate cartilage.
The optimal subset of ellipses was sought systematically, taking into account that a
uniform color and constant, relatively small structure thickness was expected.

In all the above examples, the number of elements describing an image content is
significantly smaller than the total number of pixels in that image. This may provoke
concern that a change of the image representation may lead to a crucial information loss. To prevent this, all the considered elements have some additional attributes
assigned. In the case of pixels, these attributes are their coordinates and color components. For more complex elements, the amount of information that can be assigned to
them is naturally bigger. For superpixels, this can be their center, bounding box, average color, shape descriptor, etc. In the case of ellipses, one can additionally consider
their orientation and flattening degree. Finally, for line segments, their orientation
and length as well as the characteristics of the regions on both sides of those segments

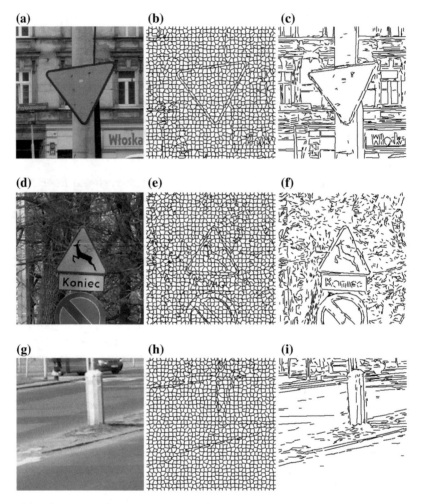

Fig. 4.1 Representation of image content: **a**, **d**, **g** sample images, **b**, **e**, **h** superpixels generated with SLIC algorithm, **c**, **f**, **i** line segments generated using LSD algorithm

can be taken into account. Another important fact is that some useful information can be also encoded in the relations between elements, for example, in the neighborhood relation. It is typical for pixels but can be also introduced for other elements. Other relations can also be defined if they are more convenient than storing the attributes of specific elements. To sum up, although the number of the elements is reduced, the information about the image content can be preserved in additional attributes of those elements and relations between them.

4.3.2 Partition

Although in active partitions the contours cannot be defined in the same way as they are in pixel representations, some partition model must be assumed to provide a feasible partition description. The most general model is one that offers a full flexibility of partition description. In that model all subsets E_O and E_B are allowed. Since the number of elements describing the image content is reduced, such an approach is acceptable in certain applications if additional constraints are imposed on the energy function and the evolution strategy. Such a model was presented in [21–23] and is used during the global analysis described further in Sect. 4.4.1. Sometimes, however, such a flexibility may be a source of problems if there is no convenient way to express expectations about the partition structure (e.g. shape) in a form of soft and hard constraints. This is illustrated in Sect. 4.4.2.

4.3.3 Evolution

Most of the typical active contour approaches use a local search algorithm as an evolution strategy. Thus, it is crucial to initialize the contour close to the desired object boundary. This constitutes one of the key problems of active contour applications. The exceptions are *Brownian strings* and *potential active contours*, which apply the *simulated annealing* algorithm as an optimization technique. And, even though other search techniques could also be used, the same approach is proposed also for active partitions, due to its simplicity and theoretical convergence with the global optimum [28–30].

In that approach, at every iteration a new solution is proposed using a solution generator G. The solution generator should generate a random solution which is close to the current one and it should enable the exploration of the whole search space during the optimization process (there should always be a possibility to generate a solution sequence transforming one solution to the other). If the generated solution is better, it is accepted as a current one. If it is worse, it is accepted with a probability depending on the difference in objective function values and on an artificial parameter, called temperature. The temperature decreases during the whole process, thus reducing also the probability of accepting worse solutions (at the beginning the temperature is selected, such that the probability of accepting worse solutions is equal to 0.8). In theory, if the whole process is sufficiently slow (infinite) this procedure guarantees that local optima are avoided. Naturally, practical applications must have a finite number of iterations, but even then the obtained results are usually satisfactory.

The choice of generator G depends on the selected partition model. In a general case, the simplest G^f generator can be considered, where the generation of a new partition involves the movement of a single element from E_O to E_B, or in an opposite direction. As shown further, such a generator, flexible as it is, does not take into account the spatial relationships between elements. Those relationships would allow the addition of topological constraints, which in turn would result in more natural partitions.

In the experiments presented in this chapter, two additional modifications were introduced to the standard *simulated annealing* algorithm. Firstly, the temperature was not decreased at every iteration but every few iterations (after a number of iterations L). The 100 temperature changes were allowed during the optimization process, since then the probability of accepting worse solution was almost equal to 0 (the exponential cooling scheme was considered with the 0.95 factor). Secondly, before every change of temperature the best solution found so far was set as the current one. Of course, since *simulated annealing* is a non-deterministic algorithm, there is a need to ensure that the results obtained are acceptable and repeatable. It can be done by proper choice of L which should be selected for a given application in its training phase.

Finally, let us explain the process of partition initialization. Here, also different strategies can be used. For example, it may be assumed that at the beginning $E_O = E$ and $E_B = \emptyset$. In the experiments presented below, other approaches were used, depending on the assumed partition model and solution generator G (initial solutions should not violate generator constraints).

4.4 Example

The present paper examines the active partition approach to object localization, focusing specifically on the problem of warning road sign detection. The problem in question is split into two phases—global and local. The global phase aims at localizing the areas of yellow color. In the local phase, those areas are analyzed in detail to find warning signs. Such an approach should correspond to fast inspection of the viewed scene to find the regions of interest and to the careful analysis of those regions. To some extent it should also imitate a conscious human-specific process of warning sign localization. In both phases, the same superpixel image representation is used. This choice was based on the assumption that human attention focuses on compact, homogenous regions rather than on single pixels. In the rest of this chapter, all the operations connected with colors are performed using the CIELab color space. In particular, it is used in the SLIC algorithm to generate superpixels and whenever the similarity of colors is discussed.

4.4.1 Global Analysis

The goal of this phase is to localize compact, yellow regions of interest. In real scenes, of course, there can be more than one such region and, naturally, not all of them need to represent warning signs. A superpixel representation should provide the ability to

Fig. 4.2 Sample input of global analysis: **a** the image with a generated superpixel representation, **b** distribution of ν_{yellow} (the brighter color, the more yellow color is present in the superpixel and its neighborhood)

(a)

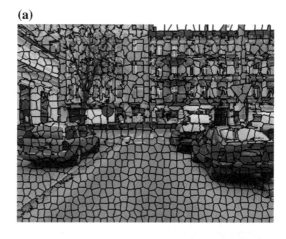

(b)

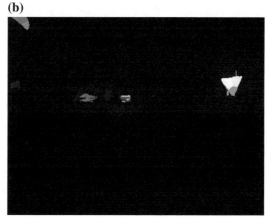

avoid problems with local color discontinuities (e.g. due to noise) at the pixel level. A sample image is presented in Fig. 4.2a to illustrate the concepts discussed in this section.

4.4.1.1   Energy

Since yellow regions are to be found, the energy evaluating the partitions should ensure that all the superpixels in E_O are to some extent yellow. This requirement is, however, not sufficient. The regions of interest may be composed of many connected superpixels and the above-mentioned requirement will be satisfied for every subset of those regions. Consequently, a natural expectation is that superpixels in E_B are not yellow. This can be expressed in the following energy function:

$$E_{color}^c(\text{P}) = w_O \sum_{e \in E_O} I(\mu_{color}(e) < t) + w_B \sum_{e \in E_B} I(\mu_{color}(e) \geq t) \tag{4.1}$$

The μ_{color} represents the percentage of pixels of a given color within a superpixel and $t = 0.4$ is an arbitrarily selected threshold. The function I returns 1 if the given condition is true, otherwise it returns 0. This objective is composed of two components. The first should ensure that, for the optimal partition, E_O contains only superpixels with a significant number of pixels of a given color. The second should minimize the number of such superpixels in E_B. Weights w_O and w_B provide the ability to control the influence of the components. If not specified otherwise, it is assumed that both of them are equal to 1.

The partition P minimizing the above, crisp energy function naturally represents the regions of interest. However, from an optimization perspective, this objective function has one drawback. If subset E_O is far from the optimal one (the distance in an image plane is considered) all the local modifications of the same size result in the same change of the energy function value. It means that there is no guidance available for the search algorithm on where the optimum is located. Thus, the *simulated annealing* requires more iterations to find a proper solution (the L value must be increased). To overcome this inconvenience, the fuzzy variant of the energy can be defined. First, the color influence for each superpixel is calculated:

$$\theta_{color}(e) = \sum_{e' \in E} \frac{\mu_{color}(e')}{1 + w\rho(e, e')} \tag{4.2}$$

Its value depends on the distance ρ between superpixel centers. Parameter w, which provides the ability to control the strength of the influence, should depend on the image size (in this work $w = 1$). Next, the obtained values are scaled to fit into the $[0, 1]$ interval:

$$v_{color}(e) = \frac{\theta_{color}(e)}{\max_{e' \in E} \theta_{color}(e')} \tag{4.3}$$

An example of v_{yellow} distribution among superpixels is depicted in Fig. 4.2b. Finally, the fuzzy energy value is computed using the following formula:

$$E_{color}^f(\text{P}) = w_O \sum_{e \in E_O} (1 - v_{color}(e)) + w_B \sum_{e \in E_B} v_{color}(e) \tag{4.4}$$

4.4.1.2 Generator

As already mentioned, the general generator G^f produces solutions that are close in a subsets space. The generated modifications, however, do not necessarily reflect local, spatial deformations of E_O in an image plane. To have this property, the solution generator should take into account the spatial relationships (neighborhood relations) between superpixels. Such a spatial relationship can be easily computed. Two superpixels can be considered neighbors if they are adjacent, i.e. if they have at least one pair of adjacent pixels. The neighborhood relationship provides the ability to define some additional topological concepts. For example, borders b (E_O) and b (E_B) can be defined as subsets of E_O and E_B where elements have at least one neighbor from E_B and E_O, respectively. Thanks to this, two additional generators, taking into account the spatial distribution of superpixels, can be defined:

- G^s—generator which either removes one element on the border b (E_O) or adds one element on the border b (E_B) (of course b (E_B) is adjacent to b (E_O)),
- G^c—generator which behaves in the same way as G^s except that it prevents new solutions from having holes or being composed of two disconnected parts (preserves connectivity).

The second generator can be useful if only connected subsets E_O are to be extracted.

Because the initial partition must not violate generator constraints, in all the experiments presented in this section a random element is selected from E to initialize the partition. Next, all the elements that are spatially close to this element in the given range are added to constitute E_O. Again, the neighborhood relation is used to decide which superpixels are close to each other. The random choice of the initial element should demonstrate that the proposed methodology helps to avoid problems with careful partition initialization.

4.4.1.3 Repeatability

The *simulated annealing* is a non-deterministic optimization algorithm. Consequently, there is a concern that this algorithm does not guarantee repeatable solutions. The concern is the more reasonable that in the presented variant of active partitions no special limitations have been imposed on the location of initial solutions.

Thus, in order to prove that the approach presented provides stable results, another experiment was conducted, aimed at selecting a proper value of L. In this experiment, for selected values of L, the partition evolution was repeated 50 times for random initial partitions. The obtained results were summarized in several ways. The distribution of final energy values is presented in Fig. 4.3a. Figure 4.3b presents the percentage of superpixels in E that always belong to either E_O or E_B. The graphical representation of repeatability is depicted in Fig. 4.3c–f. The white color indicates the pixels that are always partitioned in the same way, whereas black suggests a lower repeatability of superpixel assignment. The more intense the black, the lower the repeatability. If the evolution is repeatable, the whole image should be white. It can be observed that (L $\geq$ 50) several stable local optima are always found. To achieve perfect repeatability (global optimum) the optimization has to last longer (L $=$ 5000). Of course, those values are applicable only to the class of images under examination and the considered energy and solution generator.

Repeatability is a key issue connected with the evolutions ability to explore the whole search space, in particular if random initial solutions are allowed and local

(a)

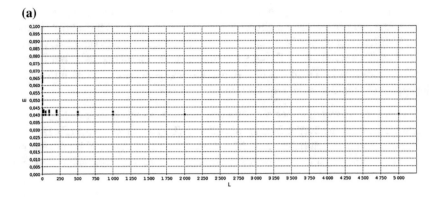

(b)

L											
1	2	5	10	20	50	100	200	500	1000	2000	5000
71.4%	96.8%	97.9%	98.1%	98.8%	99.3%	99.3%	99.3%	99.4%	99.5%	100%	100%

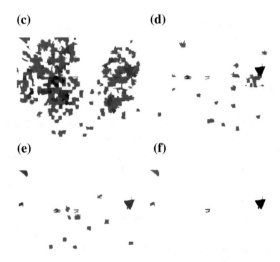

Fig. 4.3 Evolution repeatability (50 executions with random initial solutions): **a** distribution of energy E^f_{yellow} value for optimal solutions for different number of iterations L, **b** percentage of superpixels that were always assigned either to E_O or to E_B, **c, d, e, f** visualization of the assignment changes

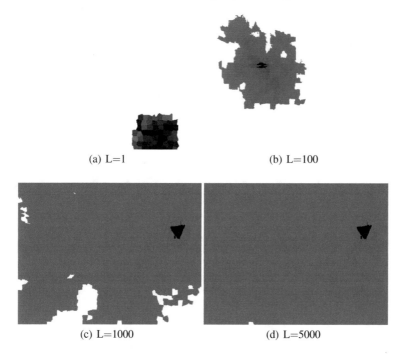

(a) L=1   (b) L=100

(c) L=1000   (d) L=5000

Fig. 4.4 Image exploration while partition evolution with G^c generator (the darker the color the more frequently a given superpixel was assigned to E_O)

partition modifications are performed by the solution generator. In Fig. 4.4 this exploration ability was presented for different L values. The more black color, the more often a given superpixel was assigned to E_O in a single run of the *simulated annealing* algorithm. This experiment also proves that for L = 5000, in the presented task, it should be possible to find a global optimum (all the superpixels were assigned to E_O at least once).

4.4.1.4 Multiple Objects

The goal of the global analysis is to quickly find an approximate position of all yellow and connected regions. For that purpose E_{yellow}^f and G^c are used. Unfortunately, active partitions, just like active contours, are naturally designed to find only single objects (usually only one optimum of the energy function is sought). Here, a simple modification was introduced to enable a multiple object localization. When the optimum is found, the energy function is modified by setting μ_{yellow} values equal to 0 for all those elements that belong to E_O. This process is repeated until no significant optimum is found. Because it does not matter in what order those optima are detected there is no need to explore the whole search space. Thus, it is assumed that L = 200. Thanks to this and the initial reduction of image size it was possible to speed up

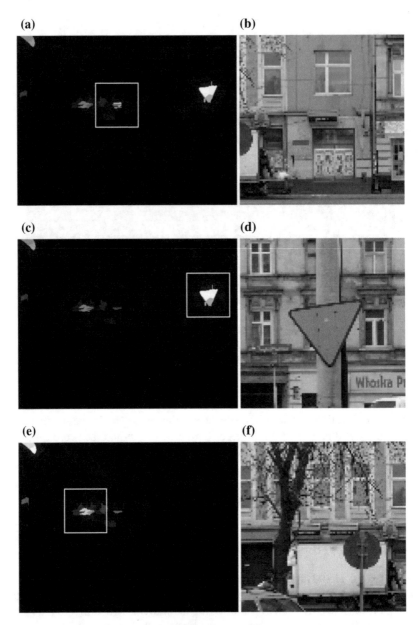

Fig. 4.5 The process of detection of multiple regions of interest: **a, c, e** changes in $\nu_{yelllow}$ distribution, **b, d, f** extracted region

the whole process. The sample results of the multiple object localization process are shown in Fig. 4.5.

4.4.2   Local Analysis

The goal of the local analysis is to find warning signs in the previously localized regions of interest. The diversity of the images leads one to assume that there are more than one yellow regions in the analyzed image. Moreover, it may happen that those regions are adjacent, as the sign and information plate in Fig. 4.1d. In such a situation, the application of the active partition technique with E^f_{yellow} and G^c cannot give satisfactory results. To overcome this problem, additional knowledge is required.

4.4.2.1   Model

A closer analysis of the results presented in Fig. 4.6 reveals that the previous approach does not take into account shape expectations. Those expectations can be added in many different ways. A typical approach would involve defining an additional energy component evaluating the similarity of the partition to the triangle. Although it is not impossible, such soft constraints are usually problematic, especially if they are supposed to be scale and rotation invariant. That is why, in the presented work, knowledge was added by changing the partition model to allow only triangles to be generated. In this model, the partition is described by three superpixels selected from E. Their centers (it is assumed that they are always organized in an anti-clockwise order) constitute a triangle. Superpixel is an element of E_O if at least one of its pixels lies inside this triangle. The rest of superpixels forms E_B.

4.4.2.2   Generator

A modified partition model requires specialized solution generators. They should modify the partition by moving triangle vertices. Two such generators are described below:

(a)   **(b)**   **(c)**

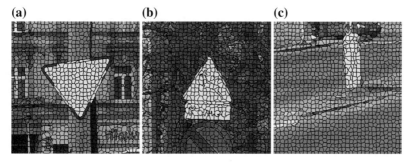

Fig. 4.6 Optimal partitions for standard partition model, E^f_{yellow} energy and G^c generator

- G^v—the generator which selects a random vertex and replaces it by one of its neighbors (the above-described neighborhood relationship is used); it forbids any modification that would change the anti-clockwise order of vertices,
- G^e—the generator which behaves in the same way as G^v but also prevents any modifications that would lead to a non-equilateral triangle (for every triangle side it checks if the corresponding height has an expected length).

The changes in the model (search space) and in the solution generators (hard constraints) entail that a feasible initial solution for the *simulated annealing* algorithm is necessary. To achieve this goal, in all the experiments the equilateral triangle of a maximum size with one horizontal side is generated (although it is not random, it still does not depend on the image content). The results obtained for those generators and E_{yellow}^f are presented in Fig. 4.8.

4.4.2.3 Energy

The above results are still not satisfactory if there are two adjacent yellow regions. This can be overcome by considering another objective function. So far, E_{yellow}^f has not taken into account the spatial distribution of colors in E_O and E_B. It is known, however, that warning signs have a red border enclosing its inner, yellow area. This observation can be expressed in the following way:

$$E^b(P) = w_O \sum_{e \in b(E_O)} (v_{red}(e) - v_{yellow}(e)) + w_B \sum_{e \in b(E_B)} (v_{yellow}(e) - v_{red}(e)) \quad (4.5)$$

On the E_O border this energy function expects yellow superpixels, not the red ones, whereas on the E_B border the expectation is exactly the opposite. Weights w_O and w_B (here equal to 1) enable the control of a trade-off between those two expectations. Sample distributions of v_{yellow} and v_{red} for image presented in Fig. 4.1d are depicted in Fig. 4.7. In Fig. 4.8c the result of triangle evolution for E^b with G^e is presented.

Fig. 4.7 Sample yellow and red color distribution

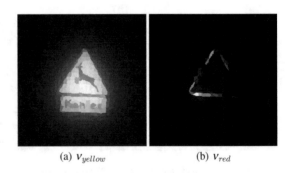

(a) v_{yellow} (b) v_{red}

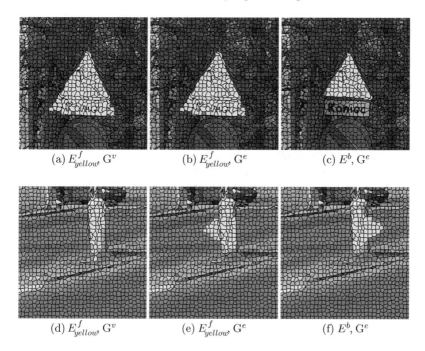

Fig. 4.8 Results of evolution for triangular partition model with different energy functions and solution generators

4.4.2.4   Missing Objects

Not all of the regions of interest must contain warning signs. The proposed active partition approach will of course work for such images and, what is more, it will generate some optimal results. Samples are shown in Fig. 4.8. Those results are reasonable as the algorithm tries to find the best position of the triangles. To automatically distinguish such cases, without the need of visual inspection of the results, the values of energy functions for optimal partitions can be analyzed. If something is wrong, these values are significantly higher than those obtained for correct structures.

4.5   Summary

The approach proposed in this chapter is a generalization of the active contour technique which can be applied to more sophisticated image content representations than raw pixel data. Its main advantage is the reduction of the search space, which enables the application of evolution strategies that are less sensitive or invariant to the choice of initial solutions (Sect. 4.4.1.3). This also means less strict assumptions about feasible objective functions. Consequently, a more natural way of expressing the expectations about the structures of interest is provided (Sects. 4.4.1.1 and 4.4.2.3).

As in the case of active contours, the knowledge required for proper analysis of the image content in active partitions can be incorporated into the search process in three ways. A proper partition model may limit the set of acceptable partitions (Sect. 4.4.2.1). The same goal may be achieved by using additional evolution constraints (region connectivity and equilateral triangles in Sects. 4.4.1.2 and 4.4.2.2, respectively). Finally, information about the expected characteristics of the partitions may be incorporated into the energy function.

Another remarkable aspect of the presented approach is its flexibility. As demonstrated by the two-stage process of warning sign detection, it can be applied to both global and local image analysis. Moreover, in the global phase, multiple objects can be localized (Sect. 4.4.1.4) by adaptive modification of the energy function.

The proposed methodology endeavours to imitate, at least to some extent, the conscious human-specific process of image analysis. Various approaches have been put forward to model the activity of the human vision system. In the literature, there are many methods that have achieved outstanding results in the field of image content understanding—*convolutional neural networks* ([31]) being a perfect example. Those models, however, are hardly interpretable and usually require huge data sets from which the expert knowledge could be automatically extracted in a training phase. Those data sets are not always easily available, especially in medical applications. That is why it may be more convenient to encode expert experience directly using the approach presented in this chapter.

As a main challenge for future work with active partitions, the choice of the best image representation should be mentioned. And this is not only the problem of optimal parameter selection (i.e. parameters of SLIC, LSD or CEC algorithms). As it was presented in this chapter, different approaches may be considered. Superpixels and ellipses focus on local region homogeneity, whereas line segments indicate some homogeneity discontinuities. These are not, however, the only possibilities, and, since all of them possess different properties, a good idea could be the fusion of those image content descriptions. Yet, the drawback of those representations is a fact that they are chosen arbitrarily for different classes of images. Perhaps a better approach would be an automatic design of such descriptors for a given object localization problem. All those aspects are under further investigation.

Acknowledgements This project has been funded with support from the National Science Centre, Republic of Poland, decision number DEC-2012/05/D/ST6/03091.

References

1. Kass, M., Witkin, A., Terzopoulos, D.: Snakes: active contour models. Int. J. Comput. Vision, 321–331 (1988)
2. Malladi, R., Sethian, J.A., Vemuri, B.C.: Shape modeling with front propagation: a level set approach. IEEE Trans. Pattern Anal. Mach. Intell. **17**(2), 158–175 (1995)

3. Casseles, V., Catte, F., Coll, T., Dibos, F.: A geometric model for active contours in image processing. Numer. Math. **66**, 1–31 (1993)
4. Caselles, V., Kimmel, R., Sapiro, G.: Geodesic active contours. Int. J. Comput. Vision **22**(1), 61–79 (1997)
5. Yezzi, A., Kichenassamy, S., Kumar, A., Olver, P., Tannenbaum, A.: A geometric snake model for segmentation of medical imagery. IEEE Trans. Med. Imaging **16**(2), 199–209 (1997)
6. Osher, S., Sethian, J.A.: Fronts propagating with curvature dependent speed: algorithms based on hamilton-jacobi formulations. J. Comput. Phys. **79**, 12–49 (1988)
7. Cootes, T.F., Taylor, C.J.: Active shape models—smart snakes. In Proceedings of 3rd British Machine Vision Conference, pp. 266–275. Springer (1992)
8. Grzeszczuk, R., Levin, D.: Brownian strings: segmenting images with stochastically deformable models. IEEE Trans. Pattern Anal. Mach. Intell. **19**(10), 1100–1113 (1997)
9. Tomczyk, A., Szczepaniak, P.S.: Adaptive potential active contours. Pattern Anal. Appl. **14**, 425–440 (2011)
10. Lu, H., Li, Y., Wang, Y., Serikawa, S., Chen, B., Chang, J.: Active contours model for image segmentation: a review. In Proceedings of the 1st International Conference on Industrial Applications Engineering, pp. 104–111. The Institute of Industrial Applications Engineers, Japan (2013)
11. He, L., Peng, Z., Everding, B., Wang, X., Han, ChY, Weiss, K.L., Wee, W.G.: A comparative study of deformable contour methods on medical image segmentation. Image Vis. Comput. **26**, 141–163 (2008)
12. Leroy, B., Herlin, I.L., Cohen, L.D.: Multi-resolution algorithms for active contour models. In Proceedings of International Conference on Analysis and Optimization of Systems, pp. 58–65 (1996)
13. Staib, L.H., Duncan, J.S.: Parametrically deformable contour models. In Proceedings of IEEE Computer Society Conference on Computer Vision and Pattern Recognition, pp. 98–103 (1989)
14. Jacob, M., Blu, T., Unser, M.: A unifying approach and interface for spline-based snakes. In Proceedings of SPIE Medical Imaging, pp. 340–347 (2001)
15. Schnabel, J., Arridge, S.: Active contour models for shape description using multiscale differential invariants. In: Pycock, D. (ed.) Proceedings of British Machine Vision Conference, pp. 197–206 (1995)
16. Denzler, J., Niemann, H.: Active rays: a new approach to contour tracking. Int. J. Comput. Inf. Technol. **4**, 9–16 (1996)
17. Horbelt, S., Dugelay, J.: Active contours for lipreading—combining snakes with templates. In: Proceedings of GRETSI Symposium on Signal and Image Processing, pp. 717–720 (1995)
18. Ivins, J., Porrill, J.: Active region models for segmenting medical images. In: IEEE International Conference on Image Processings, pp. 227–231 (1994)
19. Cootes, T.F., Edwards, G.J., Taylor, C.J.: Active appearance models. Lect. Notes Comput. Sci. **1407**, 484–500 (1998)
20. Tomczyk, A.: Adaptive potential active contours with elements of artificial intelligence. Ph.D. Thesis, Lodz University of Technology (2011) (in Polish)
21. Tomczyk, A., Szczepaniak, P.S.: Knowledge based active partition approach for heart ventricle recognition. In: 10th International Conference on Computer Recognition Systems CORES (2017) (Accepted for publication)
22. Jadczyk, M., Tomczyk, A.: Object localization using active partitions and structural description. In: Rutkowski, L., Korytkowski, M., Scherer, R., Tadeusiewicz, R., Zadeh, L.A., Zurada, J.M. (eds.) Artificial Intelligence and Soft Computing—ICAISC 2015, Part I. Lecutre Notes in Artificial Intelligence, vol. 9119, pp. 727–736. Springer (2015)
23. Tomczyk, A., Spurek, P., Podgorski, M., Misztal, K., Tabor, J.: Detection of elongated structures with hierarchical active partitions and cec-based image representation. In: Burduk, R., Jackowski, K., Kurzynski, M., Wozniak, M., Zolnierek, A. (eds.), Proceedings of the 9th International Conference on Computer Recognition Systems CORES 2015. Advances in Intelligent Systems and Computing, vol. 403, pp. 159–168. Springer International Publishing (2015)

24. Marr, D., Ullman, Sh., Poggio, T.: Vision: A Computational Investigation into the Human Representation and Processing of Visual Information. The MIT Press (2010)
25. Achanta, R., Shaji, A., Smith, K., Lucchi, A., Fua, P., Susstrunk, S.: Slic superpixels compared to state-of-the-art superpixel methods. IEEE Trans. Pattern Anal. Mach. Intell. **34**(12), 2274–2281 (2012)
26. von Gioi, R.G., Jakubowicz, J., Morel, J.M., Randall, G.: LSD: a line segment detector. Image Process. Online **2**, 35–55 (2012)
27. Tabor, J., Spurek, P.: Cross-entropy clustering. Pattern Recogn. **47**(9), 3046–3059 (2014)
28. Aarts, E., Korst, J.: Simulated Annealing and Boltzmann Machines. Wiley, Chichester (1990)
29. Kirkpatrick, S.: Optimization by simulated annealing: quantitative studies. J. Stat. Phys. **34**, 975–986 (1984)
30. Kirkpatrick, S., Gelatt Jr., C.D., Vecchi, M.P.: Optimization by simulated annealing. Science **220**(4598), 671–680 (1983)
31. LeCun, Y., Bengio, Y.: The handbook of brain theory and neural networks. Chapter Convolutional Networks for Images, Speech, and Time Series, pp. 255–258. MIT Press, Cambridge (1998)

Chapter 5
Model-Based 3D Object Recognition in RGB-D Images

Maciej Stefańczyk and Włodzimierz Kasprzak

Abstract A computational framework for 3D object recognition in RGB-D images is presented. The focus is on computer vision applications in indoor autonomous robotics, where objects need to be recognized either for the purpose of being grasped and manipulated by the robot, or where the entire scene must be recognized to allow high-level cognitive tasks to be performed. The framework integrates solutions for generic (i.e. type-based) object representation (e.g. semantic networks), trainable transformations between abstraction levels (e.g. by neural networks), reasoning under uncertain and partial data (e.g. Dynamic Bayesian Networks, Fuzzy Logic), optimized model-to-data matching (e.g. constraint optimization problems) and efficient search strategies (switching between data- and model-driven inference steps). The computational implementation of the object model and the object recognition strategy is presented in more details. Testing scenarios deal with the recognition of cups and bottles or household furniture. Conducted experiments and the chosen applications confirmed, that this approach is valid and may easily be adapted to multiple scenarios.

5.1 Introduction

With the newly available sensors that generate RGB-D images (3D point clouds and corresponding color images) of already reasonable quality, 3D image analysis methods are intensively being developed [1, 2]. A low-level processing of such data is usually a model-independent one and it leads to the creation of 3D maps of the environment (typically voxel- or surfel maps) [3–5]. In turn, the Ontology level of an agent system considered in AI operates on high-level symbolic entities like complex

M. Stefańczyk (✉) · W. Kasprzak
Institute of Control and Computation Engineering, Warsaw University
of Technology, ul. Nowowiejska 15/19, 00-665 Warsaw, Poland
e-mail: M.Stefanczyk@elka.pw.edu.pl

W. Kasprzak
e-mail: W.Kasprzak@elka.pw.edu.pl

© Springer International Publishing AG, part of Springer Nature 2018
H. Kwaśnicka and L. C. Jain (eds.), *Bridging the Semantic Gap in Image
and Video Analysis*, Intelligent Systems Reference Library 145,
https://doi.org/10.1007/978-3-319-73891-8_5

objects and actions. There is a need for a methodology and implementation for mid-level symbolic processing of 3D images that reliably closes the gap between these two representations.

The *knowledge-based* paradigm has been intensively studied for in the past, but preferably for 2-D image analysis (e.g. [6–8]) and has not yet been really considered for processing of RGB-D data.

Recently developed *Deep Neural Networks* (DNN) and *deep learning* techniques, are mostly successful for appearance-based object classification. They approximate functions, which apparently transform sensor data into numeric features either into segments or directly into object instances or classes [9, 10]. This is of importance when complex algorithms or functions need to be defined and implemented. Although the DNNs were applied to find bottom-up image transformations, the research on modelling of context information and top-down constraints in DNNs has also begun [11, 12]. Especially when 3D objects need to be recognized in a multi-object environment, it is crucial to explore physical and contextual object relations, like occlusion relations and the probability of common appearance in given environment. Graphical and stochastic models have proved suitable to handle such cases [13, 14]. It is still an open question whether neural networks techniques can deal with symbolic object-level and ontology-level concepts in order to mimic logical reasoning processes. Here, the knowledge-based approach leads straightforward to adequate solutions [15].

In particular, our focus is on basic scenarios for 3D object recognition that are explored in service and social robotics [16, 17]: human pose recognition, obstacle recognition/avoidance and grasping/manipulating of objects (Fig. 5.1).

Fig. 5.1 Basic scenarios for 3D object recognition in service and social robotics: human recognition, obstacle avoidance and object grasping

So far, 3D object recognition in RGB-D images follows a data-driven strategy and mainly identifies a particular known object. Some software packages, preferably available in the ROS (Robot Operating System) programming environment, are listed next. MOPED [18], a real-time Object Recognition and Pose Estimation system, recognizes objects by comparing point-based features (e.g. SIFT, SURF) and their geometric relationships with rigid 3D object models (defined by point clouds). LINEMOD [19] ("multi-modal templates for texture-less object detection") is detecting texture-less 3D objects located in a strongly textured background. Textured Object Detector—is based on the standard "bag of features" technique [20]. During training, in images containing different views of the object, image features are extracted and their descriptors are obtained. For each of those features, the 3D position is also stored. Transparent Object Detector [21] is a pipeline that can detect and estimate poses of transparent objects, given a point cloud model of an object. The ODUFinder system [22] can detect and recognize textured objects in typical kitchen scenes. The models for perceiving the objects to be detected and recognized can be acquired autonomously using the robot's camera as well as by loading large object catalogs into the system. Richtsfeld et al. [2] developed an effective object model learning approach based on surface grouping in RGB-D data. But still, object instances are modelled and not their generic types.

A variety of *model-based* techniques have been developed in order to recognize 3D objects from images—hierarchical models [23], among them deformable part-based models [24] and probabilistic graphical models [14] appear to be most successful. In our paper, a model-based approach is proposed that is related to the principles of above techniques.

First, we focus on the 3D object representation and modelling issue. A discussion of knowledge hierarchy levels in object recognition systems is provided in Sect. 5.2, while Sect. 5.3 deals with 3D modelling in RGB-D data. In Sect. 5.4, our framework for 3D object recognition is introduced. The system's concept and its main element— the knowledge representation techniques and inference rules—are presented here. System implementation is summarized in Sect. 5.5. The work is illustrated in Sect. 5.6 by an application of robot vision in a household environment.

5.2 Knowledge Representation Hierarchy

In this section, a review of some approaches to general 3D object representation in images is presented and our solution, suitable for RGB-D images, is given. General levels of information representation (also called *categories* of representation entities) for 3D object recognition in images are discussed.

5.2.1 Related Work

Early image analysis systems mainly used linear features and wire-frame models, so categorization was made accordingly. Marr [25] distinguished four main conceptual levels of information representation. The first one is the IMAGE—represented as an array of point intensities. Second is a PRIMAL SKETCH, which contains some basic structures extracted from the image, like edge segments, discontinuities of intensity, gradient zero crossings, etc. Based on these features a $2\frac{1}{2}$ SKETCH is created, describing the visible surfaces in terms of contours, orientation and roughly estimated depth (all expressed in the viewer coordinate system). The last level is the 3D MODEL, describing the shapes and their spatial organization in an object-centered coordinate system. An object is composed of both volumetric and surface primitives and is arranged hierarchically.

Lowe [26] proposed a slightly different categorization in his system, where instead of the $2\frac{1}{2}$ sketch 2D PERCEPTUAL GROUPINGS are used. This requires a clustering of image features, obtained in the previous step, into some consistent groups. This extension made the description more general, as virtually any feature can be used and not only linear segments like before. Lowe also added an explicit VERIFICATION STEP, connecting 3D models with low-level image features. This in turn put some restrictions on the features and model used—there must be defined a method for object back-projection onto the picture. The hypothesis-generation and -verification cycle as a basic 3D scene recognition strategy was modelled formally by Kasprzak [13] as a bi-directional syntactic-semantic derivation using an attributed structure grammar.

Data representation categories correspond to different processes transforming data from one form to another. Forsyth [27] distinguished EARLY VISION, consisting not only of basic operations like image preprocessing or edge detection, but also texture description and depth reconstruction from stereopsis or structure from motion. His MID-LEVEL VISION is responsible for clustering and segmentation, fitting objects to segments and tracking them. HIGH-LEVEL VISION is meant to be the place, where data is collected from multiple measurements. Hypotheses are generated and verified here. Object detection and recognition at this level is done using complex classifiers. Relationships between detected objects can also be described.

Gonzalez [28] made explicit definition of the processes by defining their interfaces—data types that are used on the input and output of a proces. For LOW-LEVEL PROCESSES a *picture* is used as both the input and output. Processes are elementary picture operations, like image filtering. MID-LEVEL OPERATIONS on pictures are responsible for their segmentation into consistent groups, their description reducing representation dimensionality and also the classification (or recognition) of those segment groups into individual objects. The output of this process, usually taking the form of classified objects characterized further by vectors of numbers (attributes), is supplied to the HIGH-LEVEL VISION—a symbolic processing level that is responsible for image understanding and performs cognitive reasoning.

5.2.2 Proposed RGB-D Data Hierarchy

In case of depth data, sometimes additional processing steps are required, which are somewhere in between preprocessing and feature extraction. Examples of these operations are:

- normal vector calculation or curvature estimation for a surface patch,
- conversion from a depth map to full XYZ coordinates of a point cloud,
- transformation between different coordinate systems.

As a result, it seems reasonable to put an additional DATA EXTENSION layer between the signal- and feature extraction layers. Wrapping up, data representation in our 3D object recognition system is composed as a hierarchy of 6 layers, given below.

Hardware layer. It contains actual devices for data acquisition (cameras, sensors).

Signal layer. It is responsible for image pre-processing and data preparation for feature extraction (e.g. computing edge images or labeling consistent regions). These operations need no any external information and can be run using only one (current) picture.

Extension layer. It contains processes for computation of new data representation (from those returned by the sensor) or transformation of those using some external information (like sensor position or context images). These are operations like background subtraction, normal vector calculation, depth extraction from stereo images, coordinates transformation etc. Processes from extension and signal layer can be interleaved.

Feature extraction layer. It extracts condensed, numerical information from pictures, such as feature points, edges or blob segments. The produced information may vary from simple parameters, like line end points or segment mask, through some statistical information, like mean color or surface convexity [29], to higher-level interpretation, like parameters of inscribed surfaces [30].

Object recognition layer. It gathers segments and features computed by the lower layer and composes them to form an object of interest, based on some kind of provided model (at this stage the recognition is limited to single, isolated, objects). Object recognition processes can influence the way lower level processes work (e.g. changing parameter settings of feature extraction functions in order to return more or less crisp data).

Cognitive layer. Here a higher-level (symbolic) reasoning about the scene occurs (e.g. physical and contextual relations between objects are explored). This layer is also responsible for accumulating information in time (e.g. to allow lateral processing that improves the estimation of object parameters from multiple measurements).

Particular implementations of the processes located in low- and mid-level layers depend on the chosen form (*modality*) of the object model. Such different modalities (e.g. 2D edge model, 3D surface model) require different operations on previous layers. On the other hand, an object recognition system should be generic and should allow usage of generic model description, making it possible to recognize different instances of the same kind of objects (like different sizes of jars or widths of doors).

5.3   3D Object Modelling

In this section, possible 3D object modeling modalities are discussed. The focus is on computer vision applications in indoor autonomous robotics, where objects need to be recognized either for the purpose of being grasped and manipulated by the robot, or where the entire scene must be recognized to allow high-level cognitive tasks to be performed.

Both steps require different modalities of 3D object models: geometric modelling (of physical shapes) or conceptual modelling (aggregations of parts).

5.3.1   Geometric Primitives

Grasp planners [31] work using geometric models of actual instance of the object, and this must be in a form compatible with physical engines used in a machine process, like triangular meshes or, even better, a composition of basic shapes. Methods for representation of geometric primitives can be generally divided into two main groups—DISCRETE and CONTINUOUS.

Discrete description keeps information about some finite number of elements or features, sampled from the original solid/object.

- When points are sampled from the surface of the object a POINTCLOUD is produced. The data structure of a point consists of its spatial coordinates (usually given in the Cartesian coordinate system), but it also can include other information, like the surface color or normal vector of a surface patch around this point. Points can be further expanded to surface elements (called surfels) [5], which are small surface patches approximated by discs.
- When information about volume of the object is crucial, another representation can be used, utilizing some volumetric shapes instead of points. Those elements, called VOXELS, are usually modeled with cubes, formed in either regular grid (with every element having the same size) or hierarchic structure (like octree), allowing for better approximation of complicated shapes with smaller number of elements.

Main advantage of discrete representations is the ease of model creation—in vast majority of cases depth sensors return information in form easily convertible to pointclouds. Hence, models can be built from few object views only [32]. The discrete model accuracy is proportional to the density of the pointcloud or a voxel grid, which is proportional to model size (verbosity). It must be noticed, that after the image sampling step some information about the scene is lost (the surface between sampled points).

In contrary, continuous models represent the scene information by parametric functions of continuous spatial variables, allowing recovery of any point on the object's surface. One example of such model, described in [33], is a FUNCTIONAL model, where a shape is given by an equation specifying a continuous set of points.

- PARAMETRIC equations, in form of $F : T^m \Rightarrow X^n$ explicitly define the object's points in a n-dimensional space based on m free parameters. For three-dimensional objects $n = 3$, while for $m = 1$ curves are defined and for $m = 2$—surfaces. This representation form allows to model a broad range of shapes, from simple volumes to superquadrics. The ability to directly enumerate surface points makes it easy to convert such a model to a pointcloud with theoretically unlimited precision (density).

- In contrast, if the function is given in IMPLICIT form, $F(X^n) \Rightarrow Z$, it can be treated as a characteristic function for the described shape. Surface of the object is made of all points satisfying $F(X^n) = 0$, which is sometimes hard to calculate. On the other hand, this representation makes it very easy to check, whether point lays inside the given volume by checking the condition: $F(X^n) < 0$. Hence, it can easily be converted to a voxel grid. It also allows for a simple collision detection. Calculating objects intersection is also much easier using few implicit functions and checking them one by one.

The acquisition of functional models is computationally more expensive than of discrete models, as it requires some sort of surface fitting procedure, along with constrained set of shapes [34].

Another kind of continuous models are COMBINATORIAL models, locating itself between simple and complex object representations.

- Models can be created by combining a finite set of functional submodels using set operations. Using implicit functions in a way mentioned earlier, a MEMBERSHIP object definition can be created—for example, in terms of an intersection or sum of few functions. This way more complicated shapes can be created from simpler ones, without the need to use complex equations.

- TOPOLOGICAL representation is another kind from this family. Topological structure holds the spatial relations between subelements, like two faces touching each other or two points being connected with a line. Although the elements may be of any type (e.g. parametric surfaces with values from some bounded set), the most common type of this representation is a mesh, being closely related to discrete representations. Points are connected with edges, from which planar polygons (triangles are most common) are composed. Polygons are grouped into surfaces, and those into a final shape.

5.3.2 Complex Objects

Contrary to an object grasping and manipulation application, vision systems employed for cognitive tasks require models to be composed of parts that correspond (directly or via its parts) to segments that are possible to be detected in images (e.g. feature points or textures from color images, surface patches from range data). The hierarchic nature of the model is an added value, making it possible to build complex objects out of simpler ones.

- CONSTRUCTIVE SOLID GEOMETRY (CSG) is a way to describe objects using logical operations on primitives (like sum, difference, common part) and some geometrical modifiers (expansion, morphing). Representation of the primitives must be appropriate for this task, and functional models suits here the best.
- Another method is just a simple HIERARCHICAL model, where primitives are connected with each other using joints, and the transformation between them are done using homogeneous matrices. That kind of representation makes it possible to create objects with internal degrees of freedom (like cabinet with doors), as homogeneous transformations between parts can be parameterized. Similar strategy can be used to describe whole scene as one tree with multiple smaller object trees connected [35].

Another important factor, when talking about complex objects, is the ability to define different LEVELS OF DETAIL. When observing an object from far distance only the biggest parts are visible, and those can be also simplified, as finer details may disappear. The parts of the model can be differentiated—for coarse model only some global percepts can be used for description (like histograms of colors or silhouettes), whilst closer view can incorporate local texture descriptors and good quality depth data. When the distinction is made for only two levels—coarse and fine—the first one can be used for fast generation of object hypotheses, while the second level is used for hypothesis verification.

5.4 System Framework

5.4.1 Solution Principles

Object recognition is considered to be an intermediate image analysis level, located between the low-level image segmentation processes and the ontology level of a scene understanding process.

Both main computational paradigms, the knowledge-based one (e.g. model-based) and the neural network one (e.g. appearance-based), try to overcome the limitations of available 3D computer vision systems by concentrating on three basic design principles:

1. hierarchical framework-like architecture with increasingly abstract representation levels;
2. iterative control of object recognition by integrating bottom-up, top-down and lateral processing;
3. adaptability of the general framework to particular application domain by learning the object model and recognition strategy.

Following about design principles, in this section a 3D object recognition framework is developed, which integrates several methodologies, like proposed by us earlier [23, 36]: a generic (i.e. type-based) object representation (using semantic networks), trainable transformations between abstraction levels (performed by neural networks and deep learning techniques), techniques for reasoning under uncertain and partial data (e.g. Bayesian networks and Dynamic Bayesian Networks, Fuzzy Logic), an optimized model-to-data matching (e.g. constraint satisfaction and optimization problems) and efficient search strategies (controlling alternative realizations of data-driven hypothesis generation and model-driven hypothesis verification steps).

5.4.2 Knowledge-Based Framework

Knowledge-based systems are decomposed into two main parts: the *knowledge base* and the *control* [15]. Our particular system structure is depicted in (Fig. 5.2).

The knowledge base contains three elements: the *MODEL*, the *DATA* and *inference RULES*. In this approach, the model has a hybrid form, built around the structure and inference mechanism of a *semantic network*. Besides the declarative model and

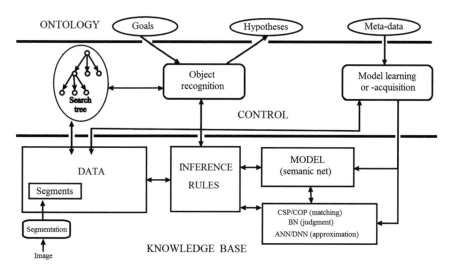

Fig. 5.2 Our knowledge-based framework

data structure expressed by the *concepts* and their interconnections in the semantic network, other techniques are here integrated: a dedicated *constraints satisfaction problem* for model-to-data matching, a *Bayesian network* for quality judgement of an *instance* or *constrained concept* and a *neural networks* (eventually *deep learning neural networks* for attribute computation.

The inference rules take the form of: "IF (condition) THEN add instance or constrained concept to DATA".

The DATA holds current symbolic descriptions of the signal (image) in form of instances and constrained concepts, generated initially by low-level image analysis (basically—image segmentation) and later as a result of the model-based inference process.

The CONTROL part performs a search in the space of competitive hypotheses, guided by their judgement values. In every step an available subset of data has to be matched with some model concept in order to satisfy the condition of some inference rules. Hence a lot of alternative decisions have to be controlled.

The model-to-data matching is seen as a specific *constraint satisfaction problem* or *constraint optimization problem*, but for many concepts it needs to be satisfied only partially (assuming a partial match).

The judgement of concept instance is estimated by a stochastic inference in a Bayesian net that is linked to given concept.

A general-purpose control strategy is defined by a space search algorithm.

5.4.3 Semantic Net

Common to semantic networks is the explicit structuring of domain knowledge along two hierarchies: the decomposition (vertical) hierarchy and the specialization (horizontal) hierarchy of concepts.

Starting from the pixel level the vertical hierarchy expresses increasingly abstract representation levels ("part" or "concrete" links). Simple elements are combined into more complex one, being parts of objects and scenes. Specialization links ("spec") represent inheritance relations between elements at the same abstraction level.

Every node (called "concept") represents some object category and it contains a parameter vector (called "attributes"), where every parameter is evaluated by some *term*, and every concept defines a set of constraints, evaluated by *predicates*, among its parts and related concepts.

A procedural part is added to the semantic network that implements the semantics of terms and predicates. It consists of functions for attributes and relations for predicates. In fact, a semantic network is an object-oriented form of a specific predicate logic. If we allow concept attributes to hold default values then such a semantic network represents a non-monotonic logic.

The part- and spec-links have an appropriate representation in logic. The relation, "{set of parts} $-part->$ concept C", is equivalent to a formula built around the implication symbol, in straight direction, $(C_{part1} \wedge C_{part2} \wedge \ldots \wedge C_{partN} \Rightarrow C)$, and in the reverse direction, $(\forall_{I \in 1,\ldots,N}(C \Rightarrow C_{partI}))$.

Similarly, the dependence, "base concept $-spec->$ inherited concept", is equivalent to a formula: $C_{inherited} \Rightarrow C_{base}$

5.4.4   Bayesian Net

A Bayesian net (BN) is a simple, graphical notation for conditional independence assertions and hence for compact specification of full joint distributions: (1) a set of nodes, one per stochastic variable; (2) a directed, acyclic graph (link means "direct influence")—incoming links of given node represent a conditional distribution for this node given its parents, $P(X_i | Parents(X_i))$. In the simplest discrete case, conditional distribution is represented as a conditional probability table (CPT), giving the distribution over X_i for each combination of parent values.

An illustration of a Bayesian net is shown in (Fig. 5.3)—it represents variables related to a Rubik_cube concept. An intermediate level in the model represents visible faces. The lowest-level concepts represent 9 color squares, that define the texture of a face. There are also evidence nodes that represent constraints between faces (fA, fB) and constraints between squares (A, B, D).

The score of a partial solution (assignment in terms of CSP), in which some variables X_i have already been assigned to image segments l_k but not all of them, is obtained due to stochastic inference in Bayesian net. For example the computation of posterior probability of a "cube" instance (that is a *cause* in terms of BN) given its parts (that are *evidences* in BN). For example, if segments are assigned to X_0 and X_1 then one need to compute the probability: $P(cube | X_0 = l_1, X_1 = l_2)$.

Fig. 5.3 A Bayesian net structure for concept: "Rubik_cube"

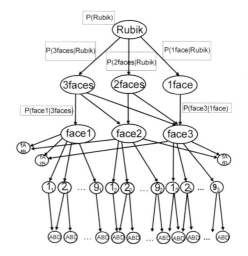

This leads to a summation of pdf over all domain values for remaining (non-evidence) variables, $X_2, ..., X_l$. Thus, scores of partial matches or a complete match, between image segments and model entities, are naturally obtained by the same evaluation method.

5.4.5 The Basic Control

The object recognition process is performed for given set of object concepts, called the GOALS G. This set can contain "concepts", "constrained concepts" or even "instances". Let M denotes concepts stored in the model base, while D is the current data set. In every single step, the basic control algorithm activates one of the five available inference rules, RULE_1, ... , RULE_5, for selected model concepts and data instances.

1. IF $G \neq \emptyset$ THEN perform a top-down goal concepts expansion (propagation of constraints), using inference RULE 4; ELSE perform a bottom-up hypothesis generation for concepts in M, based on important image segments in D, using RULE 5.
2. A bottom-up generation of "partial instances", that match the existing "constrained concepts" for obligatory parts of some modality of the selected model concept with the data instances (using RULE 1): $I_p(k) \leftarrow \{(part_i \in M_k; I_i \in D) | i \in oblig(M_k)\}$; where attributes of every instance $I_p(i)$ are $a = (S_k, R_k, t_k)$ (shape, rotation, translation);
3. Hypothesis verification: FOR every hypothesis $I_p(k)$ DO

   - constrain its remaining (non-obligatory) parts ("top-down" RULE 2) and match these parts with DATA: $I_e(k) \leftarrow \{(part_j \in M_k; I_j \in D) | j \in optional (M_k)\}$
   - Verify the hypothesis $I(k) \leftarrow (I_p(i) \bigcup I_e(k))$—create a "full instance" and re-compute its attributes a' (a "bottom-up" RULE 3).

4. Return the lattice of verified hypotheses, i.e. a graph where nodes represent hypotheses and arc—relations of mutual exclusion.

It depends on a particular search strategy (and current data and hypotheses) which step is selected and performed next.

5.5 System Implementation

The particular data types and predicates will be discussed that are implementations of nodes of the abstract semantic net (concepts) and the constraints between parts of a concept.

Two basic building blocks of the knowledge base, the MODEL and the DATA, are connected with two views of a 3D object. First is the "idealized" view, i.e. the object's type. The other one is the instance hypothesis, i.e. a set of parts (e.g. segments) recognized as an object of interest.

5.5.1 Model Structure

The *model M* is the "idealized" view of the object, describing its generic properties and allowing to recognize multiple realisations (instances) of this type of objects, like chairs of different sizes or different bottles, as long as they share some common features. A single object's model is the implementation of a dedicated *concept* from a semantic network. A model is built from PARTS P, constituting observable objects itself, CONSTRAINTS C, defining relations between those parts, ATTRIBUTES A, allowing a differentiation between instances and SCORE—a judgment of instance quality. Thus, a model is a tuple consisting of following entities:

$$M = \left\{ P = \{p_1 \ldots p_n\}, C = \{c_1 \ldots c_m\}, A = \{a_1 \ldots a_k\}, score = \{s_1 \ldots s_l\} \right\}$$
(5.1)

Parts have a pre-defined unique *role* in the model (like left leg or mug handle), while the constraints are expressed by relations between parts of a concept and are evaluated on attributes of these parts. Alternative "specialized" versions of a concept or alternative subsets of the parts of a concept (called as *modalities*), are illustrated by OR links on the diagram on Fig. 5.4.

The basic structure of a single model is represented by a graph on Fig. 5.5. Following sections describe every element of this diagram in details.

As a simple example, illustrating presented concepts, the mug object is used. Putting term *mug* into the image search engine yields a list different pictures (Fig. 5.6), but all of the presented objects possess some common elements. Every mug has a more or less toroidal handle and a main cylindrical part for liquid. They differ in size and color, but can be described using one generic model.

5.5.1.1 Parts

Part is some observable element of the object. It can be sometimes identified as a physical element, like the leg of a chair or a door knob, but in other cases it can be a more abstract thing, like the edge of a box or even a single point (or feature point) extracted from an object surface.

For each part p_i there must be assigned a CLASS ($class(p_i)$), i.e. another model representing this entities type. This defines, what kind of part it is, and whether it can be for example matched to a *cylinder* in applications, where one can observe 3D geometrical shapes, or matched with a *line* when edge-based processing is used, etc.

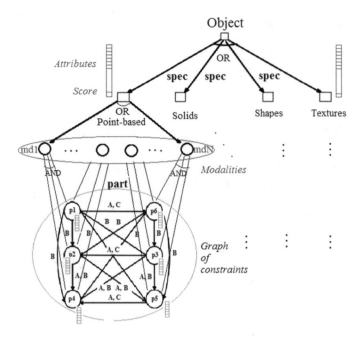

Fig. 5.4 Alternative 3D object models (specialized concepts of the "Object" concept) and typical structure of a single model

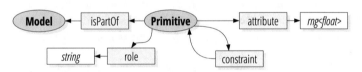

Fig. 5.5 The TBox class structure—abstract concept implementation

There can be, of course, many parts of the same class, like there are for example four legs in a chair.

To differentiate between parts of the model, each has to have defined a unique ROLE, which will be used as an identifier in further processing steps. The role of the part can be either a more abstract one, like a *left-edge* when describing geometrical shapes, or it can mimic part affordances, like a *handle* for the toroidal part and *body* for the cylindrical part of the mug model.

5.5.1.2   Attributes

Attributes describe properties of a model. In the model, an attribute is defined by its data type and the range of its allowed values. Only instances of such model with attribute values lying in given range can be considered as its proper instances. The

Fig. 5.6 Sample mugs retrieved by web image search

attribute calculation function itself can return any value from the domain D of the attribute's type:

$$Attribute(M) \in \mathcal{D} \qquad (5.2)$$

The same attribute type can be used for multiple parts, but for each of them different range of possible values can be set. For example, color of the main part of the mug (modeled as a hue component) can be set to red, while the handle can be white. Another group of attributes are of geometrical nature. Typical mugs have radius in the range from 3 to 6 cm.

5.5.1.3   Constraints

In contrast to attributes, constraints are defined on some subset (at least with two elements) of parts, and they represent some relation between them. There could be logical constraints (like checking, if some parts have the same size), spatial ones (like checking, if two lines are parallel) or others.

Fuzzy set functions for constraint evaluation return values from the $[0\ldots 1]$ range (instead of the Boolean values $\{True, False\}$), where 1 means full constraint satisfaction and 0 total inconsistency of given set of parts with examined relation. It enables to treat the result of constraint satisfaction check as an intermediate score in further processing steps, giving finally an overall score of the model's instance.

$$Predicate(p_1, \ldots, p_k) \in [0\ldots 1] \qquad (5.3)$$

For the simple mug model one can require that its handle intersects with the main part. The *intersects* constraint can be defined as a function returning 1 if the handle's (toroid) center lies on the surface of the main cylinder and gradually dropping to 0 when the toroid's center is farther than its radius from the main part. This function may be based on the distance between center of the toroid (with radius r) and the axis of the cylinder (with radius R). It looks like the one presented on Fig. 5.7. The final model structure for the mug is presented on Fig. 5.8.

5.5.2 Object Instances

A model of some object is a generic representation of its structure. When an observation is made, multiple segments can be extracted from it, and those can be classified as instances of some basic concepts (model entities), called as the *primitives* of symbolic representation. During the object recognition process, some of these primitives can be assigned to model parts (if their attributes are in desired ranges), and after satisfying the constraints of given model, they eventually lead to the creation of this model's instance (e.g. an object hypothesis).

Fig. 5.7 Sample calculation functions for the *intersects* constraint

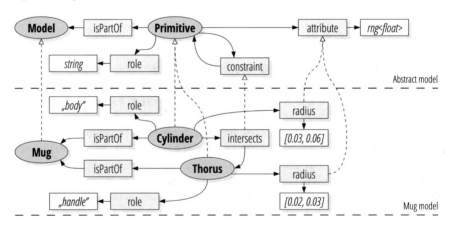

Fig. 5.8 A TBox representation of the "mug" model

5.5.2.1 Model-to-Data Matching

Each assignment of existing instances to model parts, made in a way that for each part of the model exactly one instance of proper class is selected, is called hypothesis $H(M)$. It can be defined as:

$$H(M) : \forall i \in \{1..n\} \exists j \in \{1..k\} : p_i \leftarrow inst_j \land class(p_i) = class(inst_j) \quad (5.4)$$

where n is the number of parts in the model and k is the number of already recognized instances ($inst_k$, e.g. segments). Overall hypothesis score is calculated by taking the product of all constraints for given assignment.

$$score\big(H(M)\big) = \prod_m eval(c_m, H) \quad (5.5)$$

The most naive way of generating hypotheses is to perform an *exhaustive search* in the entire space of possibilities, i.e. generating all variations of existing instances that match the model structure. This way, the number of generated hypotheses, that must be further checked for their scores, may be big or even huge:

$$|O| = \prod_{d \in D} |d| \quad (5.6)$$

where O is a set of object hypotheses and $D = \{d_i\}$ is a domain for particular part (i.e. set of all instances of the same class as given part). For each hypothesis its score is evaluated and the best ones are returned.

5.5.2.2 Matching by CSP/COP

To avoid full expanding of hypothesis before its verification (which is computationally very expensive), the matching problem can be treated as a *constraint optimization problem* (COP). Basic backtracking algorithm is used to build hypothesis step by step. After assigning a new variable (in our case assigning existing part-type instance from DATA to a yet unassigned model part) the hypothesis score is calculated and, if the score is lower than some existing threshold, current branch is pruned and the algorithm goes backward to search for other, better possibilities.

Classic CSP works until it finds the first solution satisfying all the constraints. As the score in our system can be anywhere between 0 and 1, we can compare two hypotheses and select the better one. CSP implementation is thus modified as follows. We keep track of N best hypotheses found so far (this list is empty when the search algorithm starts). At each step, the hypothesis score is calculated and, if it is lower than the worst from the list, this search tree branch is pruned. If a complete hypothesis is generated and its score is high enough, it is placed on the list. This way we achieve the same effect as in exhaustive search, but with much better performance—a lot of hypotheses are rejected at early stage.

For efficiency reasons, this basic strategy is additionally supported by three techniques: selecting the most constrained variable first, selecting the highest-scoring value first, and making a forward check of the constraints.

5.6 Testing Scenarios

As an simple illustration, generic mug model will be used (similar to presented in previous sections), with two distinct modalities of incoming data: RGB image and RGB-D image. In first scenario, edge-based analysis [37] is applied, with two basic concepts—*Cylinder* and *Arc*, corresponding to two mug parts—*body* and *handle*. Second scenario uses depth data [38], and two basic surface concepts—*Cylinder* and *Thorus*.

5.6.1 Data Acquisition

As an input data, *Complex scene 3* from WUT Visual Perception Dataset [39] was used, which contains a recorded trajectory (77 points) "around the table". The set was acquired using Microsoft Kinect sensor, and it contains, for every recorded position, a pair of images, aligned with each other: RGB image (Fig. 5.9a) and depth map (Fig. 5.9b). The selected scene contains three cylindrical objects—two mugs and one coffee jar, as well as some other kitchen utensils.

The data was acquired with hand-held sensor, thus there is no ground-truth position data and the trajectory was recovered using visual odometry solution [5].

(a)   **(b)**

Fig. 5.9 Test scene: **a** RGB image, **b** depth map

(a)   **(b)**

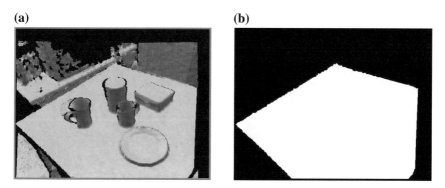

Fig. 5.10 Results of data extension: **a** calculated normal map; **b** mask of interesting scene part

5.6.2 Data Preprocessing and Extension

During preprocessing and extension phase, there are two steps worth mentioning.
One is the calculation of surface normals. For every valid point in depth map, if it
contains sufficiently big surrounding, it is used to calculate a vector perpendicular to
the surface in given point (Fig. 5.10a). Another operation is mask generation. Based
on information from control subsystem, search space can be restricted to a smaller
area—in this case only objects on the brown table are interesting for us (Fig. 5.10b)

5.6.3 Segmentation

Edge-based analysis uses a two-step segmentation process [37]. At first, only linear segments are detected (arcs and lines), and then those are connected into more
complex structures (cylinders in described scenario). To create those complex structures, the same hypothesis generation step is used, with a model describing cylinder
appearance. Final segmentation result looks like shown on Fig. 5.11a.

Surface segmentation uses RanSaC to inscribe cylindrical and thoroidal surfaces
into acquired 3D image (using points position in space as well as their normal vectors). Sample segmentation result for one view is shown on Fig. 5.11b.

5.6.4 Hypothesis Generation

Based on detected segments, initial mug hypotheses are generated for the example
model containing two parts (*body* and *handle*) and one constraint between them
(*near*). In edge based scenario, four hypotheses were generated (Fig. 5.12a). One
proper hypothesis for the mug on the right-hand side, two competing correct hypothe-

(a)   **(b)**

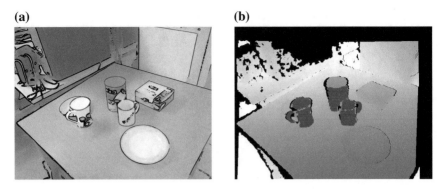

Fig. 5.11 Basic instances detected in segmentation step: **a** edges—cylinders (red) and arcs (black), **b** surfaces—cylinders (red) and thoroids (blue)

(a)   **(b)**

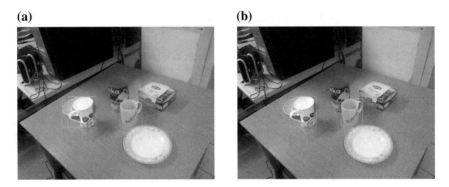

Fig. 5.12 Initial hypotheses: **a** edge-based, **b** surface-based

ses for the left-hand side mug (two different arcs for the *handle*) and one false hypothesis for the same mug (with plate taken as an *handle* part).

In second scenario, where surface-based analysis is used, only two hypotheses are generated, one for each mug on the scene. It is worth mentioning, that in both cases coffee-jar was not taken into account as candidate object because of lack of nearby *handle*-like segment.

5.6.5 Hypothesis Update and Verification

Initial hypotheses are tracked in consecutive images. When a new measurement comes in, apart from detecting new hypotheses, existing ones are matched against detected segments and their scores are recalculated. In edge-based scenario, this tracking step allows to detect false mug-plate hypothesis, as in subsequent views the underlying parts (cylinder and arc) move away from each other (Fig. 5.13a).

(a) (b)

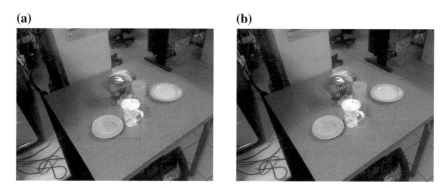

Fig. 5.13 Tracked hypotheses, green—good, red—bad: **a** edge-based, **b** surface-based

Second scenario was initialized with two proper hypotheses, so nothing is dropped in subsequent views. More measurements, however, are used to make better estimation of object parameters, for example position in space or size (using Kalman filter).

Detected (and updated) objects are returned to control subsystem for each incoming measurement.

5.6.6 Method Vulnerabilities

The proposed object recognition method is generic as it explores a generic object model. Its efficiency depends highly on the quality of particular domain model given as its input. It should be noted, that for poorly created models the recognition results may be disappointing, or the processing time can be very long. If the model is created with very high level of details and precision, e.g. every model element has highly limited range of accepted attribute values, even small sensor inaccuracies can lead to a rejection of all available image segments, resulting in a failure of object detection. On the other hand, when the model is defined with small number of constraints and very relaxed attribute restrictions, almost every image segment must be checked when building object hypotheses. This requires to expand many, if not all, possible assignments, which in turn leads to long processing times. Thus, it is important to select proper subsets of object features, small enough to keep a generic model, yet specific enough to be discriminative. Our current focus is on automated methods of creating such optimal models.

5.7 Conclusions

An application-independent generic model-based framework for object recognition in RGB-D images has been presented and verified on robot vision scenarios. It has clear advantages over existing, mostly data-driven and appearance-based approaches to object instance re-detection. First, it allows to identify what kind of knowledge is needed and to utilize existing meta-level knowledge (e.g. types of predicates and attributes commonly used for object description) to perform machine learning of model concepts (to learn concept types instead of memorizing individual instances). Secondly, common parts of object recognition systems can be pre-implemented, which increases the efficiency of system design and its implementation in different applications.

Another important advantage of proposed system is its human-oriented approach to object modelling. The decomposition of an object into simpler elements, named parts, makes it easier to further analyze the model. Using fuzzy constraint functions enables the user to focus on overall model creation, instead of performing a hand-made tuning of parameters.

Conducted experiments (described in Sect. 5.6) and the chosen applications (presented on Fig. 5.1) confirmed, that this approach is valid and may easily be adapted to multiple scenarios. In the article, we selected one example application of the system, simple enough to be easy to follow by the reader, yet covering two most popular data modalities—color images and depth measurements. Developed algorithms are independent on the data source, so exactly the same methods and algorithms can be used regardless of selected sensor. In fact, different data sources were tested (single cameras, stereo pairs, structured light), mounted in different spots (stand alone over the workbench, mounted on robots head or at the end of the arm, near the gripper).

Acknowledgements The manuscript preparation was supported by statutory funds of the Warsaw University of Technology.

References

1. Ren, X., Fox, D., Konolige, K.: Change their perception: RGB-D for 3-D modeling and recognition. IEEE Robot. Autom. Mag. **20**(4), 49–59 (2013)
2. Richtsfeld, A., Mörwald, T., Prankl, J., Zillich, M., Vincze, M.: Learning of perceptual grouping for object segmentation on RGB-D data. J. Vis. Commun. Image Represent. **25**(1), 64–73 (2014)
3. Henry, P., Krainin, M., Herbst, E., Ren, X., Fox, D.: RGB-D mapping: using kinect-style depth cameras for dense 3D modeling of indoor environments. Int. J. Rob. Res. **31**(5), 647–663 (2012). https://doi.org/10.1177/0278364911434148
4. Newcombe, R.A., Davison, A.J., Izadi, S., Kohli, P., Hilliges, O., Shotton, J., Molyneaux, D., Hodges, S., Kim, D., Fitzgibbon, A.: KinectFusion: real-time dense surface mapping and tracking. In: 2011 10th IEEE International Symposium on Mixed and Augmented Reality (ISMAR), pp. 127–136. IEEE (2011)

5. Wilkowski, A., Kornuta, T., Stefańczyk, M., Kasprzak, W.: Efficient generation of 3D surfel maps using RGB-D sensors. Int. J. Appl. Math. Comput. Sci. (AMCS) **26**(1), 99–122 (2016). https://doi.org/10.1515/amcs-2016-0007
6. Hanson, A., Riseman, E.: The VISIONS image-understanding system. Adv. Comput. Vis. **1**, 1–114 (1988)
7. Hwang, V.S.S., Davis, L.S., Matsuyama, T.: Hypothesis integration in image understanding systems. Comput. Vis. Graph. Image Process. **36**(2–3), 321–371 (1986)
8. Niemann, H., Sagerer, G.F., Schroder, S., Kummert, F.: ERNEST: a semantic network system for pattern understanding. IEEE Trans. Pattern Anal. Mach. Intell. **12**(9), 883–905 (1990)
9. Krizhevsky, A., Sutskever, I., Hinton, G.: Imagenet classification with deep convolutional neural networks. Adv. Neural Inf. Process. Syst. **25**, 1097–1105 (2012)
10. Socher, R., Huval, B., Bhat, B., Manning, C., Ng, A.: Convolutional-recursive deep learning for 3D object classification. Adv. Neural Inf. Process. Syst. **25**, 656–664 (2012)
11. Behnke, S.: Hierarchical Neural Networks for Image Interpretation. Lecture Notes in Computer Science, vol. 2766. Springer, Berlin (2003)
12. Lin, D., Fidler, S., Urtasun, R.: Holistic scene understanding for 3d object detection with rgbd cameras. In: IEEE International Conference on Computer Vision, ICCV 2013, Sydney, Australia, December 1–8, 2013, pp. 1417–1424. IEEE Computer Society, ISBN 978-1-4799-2839-2 (2013)
13. Kasprzak, W.: A linguistic approach to 3-D object recognition. Comput. Graph. **11**(4), 427–443 (1987)
14. Koller, D., Friedman, N.: Probabilistic Graphical Models: Principles and Techniques. MIT Press, Cambridge (2009)
15. Russel, S., Norvig, P.: Artificial Intelligence. A Modern Approach, 3rd edn. Prentice Hall, Upper Saddle River (2011)
16. Tsardoulias, E., Zieliński, C., Kasprzak, W., Reppou, S.: Merging robotics and AAL ontologies: the RAPP methodology. In: Progress in Automation, Robotics and Measuring Techniques. Advances in Intelligent Systems and Computing, vol. 351, pp. 285–297. Springer International Publishing (2015)
17. Zieliński, C., et al.: Variable structure robot control systems–the RAPP approach. Robot. Auton. Syst. **94**, 226–244 (2017). https://doi.org/10.1016/j.robot.2017.05.002
18. Collet, A., Martinez, M., Srinivasa, S.S.: The MOPED framework: object recognition and pose estimation for manipulation. Int. J. Robot. Res. **30**(10), 1284–1306 (2011)
19. Hinterstoisser, S., Holzer, S., Cagniart, C., Ilic, S., Konolige, K., Navab, N., Lepetit, V.: Multi-modal templates for real-time detection of texture-less objects in heavily cluttered scenes. In: 2011 IEEE International Conference on Computer Vision (ICCV), pp. 858–865. IEEE (2011)
20. O'Hara, S., Draper, B.A.: Introduction to the bag of features paradigm for image classification and retrieval (2011). arXiv preprint arXiv:1101.3354
21. Lysenkov, I., Rabaud, V.: Pose estimation of rigid transparent objects in transparent clutter. In: 2013 IEEE International Conference on Robotics and Automation (ICRA), pp. 162–169. IEEE (2013)
22. Pangercic, D., Haltakov, V., Beetz, M.: Fast and robust object detection in household environments using vocabulary trees with sift descriptors. In: IEEE/RSJ International Conference on Intelligent Robots and Systems (IROS), Workshop on Active Semantic Perception and Object Search in the Real World, San Francisco, CA, USA. Citeseer (2011)
23. Kasprzak, W., Kornuta, T., Zieliński, C.: A virtual receptor in a robot control framework. In: Szewczyk, R., Zieliński, C., Kaliczyńska, M. (eds.) Recent Advances in Automation, Robotics and Measuring Techniques. Advances in Intelligent Systems and Computing (AISC), vol. 267, pp. 399–408. Springer, Berlin (2014)
24. Felzenszwalb, P., Girshick, R., McAllester, D., Ramanan, D.: Object detection with discriminatively trained part based models. IEEE Trans. Pattern Anal. Mach. Intell. **32**(9), 1627–1645 (2010)
25. Marr, D.: Vision: A Computational Investigation into the Human Representation and Processing of Visual Information. Henry Holt and Co., Inc., New York (1982)

26. Lowe, D.G.: Three-dimensional object recognition from single two-dimensional images. Artif. Intell. **31**(3), 355–395 (1987)
27. Forsyth, D.A., Ponce, J.: Computer Vision: A Modern Approach. Prentice Hall Professional Technical Reference (2002)
28. Gonzalez, R.C., Woods, R.E.: Digital Image Processing. Prentice Hall, Upper Saddle River (2002)
29. Stefańczyk, M., Kasprzak, W.: Multimodal segmentation of dense depth maps and associated color information. In: Bolc, L., Tadeusiewicz, R., Chmielewski, L., Wojciechowski, K. (eds.) Proceedings of the International Conference on Computer Vision and Graphics. Lecture Notes in Computer Science, vol. 7594, pp. 626–632. Springer, Berlin (2012)
30. Richtsfeld, A., Morwald, T., Prankl, J., Zillich, M., Vincze, M.: Segmentation of unknown objects in indoor environments. In: 2012 IEEE/RSJ International Conference on Intelligent Robots and Systems (IROS), pp. 4791–4796. IEEE (2012)
31. Miller, A.T., Allen, P.K.: Graspit! a versatile simulator for robotic grasping. IEEE Robot. Autom. Mag. **11**(4), 110–122 (2004)
32. Łępicka, M., Kornuta, T., Stefańczyk, M.: Utilization of colour in ICP-based point cloud registration. In: Proceedings of the 9th International Conference on Computer Recognition Systems CORES 2015. Advances in Intelligent Systems and Computing, pp. 821–830. Springer, Berlin (2016)
33. Naylor, B.: Computational representations of geometry. In: Representations of Geometry for Computer Graphics, SIGGRAPH '94 Course Notes (1994)
34. Jaklic, A., Leonardis, A., Solina, F.: Segmentation and Recovery of Superquadrics, vol. 20. Springer Science & Business Media (2013)
35. Foote, T.: tf: the transform library. In: 2013 IEEE International Conference on Technologies for Practical Robot Applications (TePRA), pp. 1–6. IEEE (2013)
36. Kasprzak, W.: Integration of different computational models in a computer vision framework. In: 2010 International Conference on Computer Information Systems and Industrial Management Applications (CISIM), pp. 13–18 (2010). https://doi.org/10.1109/CISIM.2010.5643697
37. Stefańczyk, M., Pietruch, R.: Hypothesis generation in generic, model-based object recognition system. In: Szewczyk, R., Zieliński, C., Kaliczyńska, M. (eds.) Recent Advances in Automation, Robotics and Measuring Techniques. Advances in Intelligent Systems and Computing (AISC), vol. 440, pp. 717–727. Springer, Berlin (2016). https://doi.org/10.1007/978-3-319-29357-8_62
38. Wilkowski, A., Stefańczyk, M.: Detection and recognition of compound 3D models by hypothesis generation. In: Szewczyk, R., Zieliński, C., Kaliczyńska, M. (eds.) Recent Advances in Automation, Robotics and Measuring Techniques. Advances in Intelligent Systems and Computing (AISC), vol. 440, pp. 659–668. Springer, Berlin (2016). https://doi.org/10.1007/978-3-319-29357-8_57
39. Stefańczyk, M., Laszkowski, M., Kornuta, T.: WUT visual perception dataset-a dataset for registration and recognition of objects. In: Challenges in Automation, Robotics and Measurement Techniques. Advances in Intelligent Systems and Computing (AISC), vol. 440, pp. 635–645. Springer, Berlin (2016)

Chapter 6
Ontology-Based Structured Video Annotation for Content-Based Video Retrieval via Spatiotemporal Reasoning

Leslie F. Sikos

The constantly increasing popularity and ubiquity of videos urges efficient automated mechanisms for processing video contents, which is a big challenge due to the huge gap between what software agents can obtain from signal processing and what humans can comprehend based on cognition, knowledge, and experience. Automatically extracted low-level video features typically do not correspond to concepts, persons, and events depicted in videos. To narrow the Semantic Gap, the depicted concepts and their spatial relations can be described in a machine-interpretable form using formal definitions from structured data resources. Rule-based mechanisms are efficient in describing the temporal information of actions and video events. The fusion of these structured descriptions with textual and audio descriptors is suitable for the machine-interpretable spatiotemporal annotation of complex video scenes. The resulting structured video annotations can be efficiently queried manually or programmatically, and can be used in scene interpretation, video understanding, and content-based video retrieval.

6.1 The Limitations of Video Metadata and Feature Descriptors

Common *technical metadata* implemented in video files, such as duration, frame width and height, and frame rate, and *rights metadata*, such as licensing, do not convey information about the visual content and the meaning of video scenes. In fact, even *descriptive metadata*, which is the closest metadata type to content description, provide information such as title, keywords, and genre only, which is rather limited in describing audiovisual contents.

L. F. Sikos (✉)
Flinders University, Adelaide, Australia
e-mail: leslie.sikos@flinders.edu.au

© Springer International Publishing AG, part of Springer Nature 2018
H. Kwaśnicka and L. C. Jain (eds.), *Bridging the Semantic Gap in Image and Video Analysis*, Intelligent Systems Reference Library 145,
https://doi.org/10.1007/978-3-319-73891-8_6

Low-level features, such as loudness and motion trajectory, which are automatically extracted from audio and video signals, provide information that might be useful for video classification, object matching with a reference object (even if the object has been rotated and/or scaled) and object tracking. However, they are not suitable for efficient scene interpretation and video understanding, because they, similar to video metadata, do not correspond directly to the depicted concepts and events.

6.1.1   Core Video Metadata Standards

In parallel with the tremendously increasing number of online videos, many technical specifications and standards have been introduced to store technical details, and describe the features of, video resources. Beyond the proprietary tags embedded in multimedia files, multimedia metadata specifications have been standardized over the years for generic multimedia metadata, and the spatial, temporal, and spatiotemporal annotation of videos. *MPEG-7* (ISO/IEC 15938)[1] provides XML metadata to be attached to the timecode of MPEG-1, MPEG-2, and MPEG-4 contents, such as synchronized lyrics to music videos. *MPEG-21* (ISO/IEC 21000)[2] provides machine-readable licensing information for MPEG contents in XML. *TV-Anytime* (ETSI TS 102 822)[3] was designed for the controlled delivery of personalized multimedia content to consumer platforms.

The properties of some standard general-purpose metadata specifications can also be used for videos. A prime example is *Dublin Core* (ISO 15836-2009).[4] It provides, among others, descriptive metadata, such as the title and language of videos files and physical resources (DVDs and Blu-ray discs), technical metadata, such as format, and rights metadata (licensing).

Although useful, none of these metadata standards can formally describe the actual visual content of videos and the meaning of video scenes.

6.1.2   Feature Extraction for Concept Mapping

Low-level descriptors describe automatically extractable low-level image, audio, and video features, which correspond to local and global characteristics of audio and video signals, such as frequency, amplitude modulation, and motion vectors. Based on these low-level features, feature aggregates and statistics can be computed, including various histograms, which can be used in scale-invariant object recognition, object tracking, and action recognition. Such descriptors, many of which are defined

---

[1] https://www.iso.org/standard/34230.html.

[2] https://www.iso.org/standard/35367.html.

[3] http://www.etsi.org/technologies-clusters/technologies/broadcast/tv-anytime.

[4] https://www.iso.org/standard/52142.html.

in the aforementioned MPEG-7 standard, are suitable for the numeric representation of audio waveforms and video signals.

Common *visual descriptors* capture perceptual features, such as color, texture, and shape, which can be useful for frame-level video annotation, and motion, which corresponds to camera movements and moving objects. Examples include the *dominant color descriptor*, which describes dominant colors, the *homogeneous texture descriptor*, which characterizes regional textures, the *region-based shape descriptor*, which represents pixel regions that constitute a shape, the *camera motion descriptor*, which describes global motion parameters with professional video camera movement terms, and the *motion trajectory descriptor*, which captures the displacement of objects over time using spatiotemporal localization with positions relative to a reference point that are described as a list of vectors. The audio channel of video files can be described with temporal, spectral, cepstral, and perceptual audio descriptors, such as the *zero crossing rate descriptor*, which is suitable for determining whether the audio content is speech or music, the *spectral moments descriptor*, which is useful for determining sound brightness and music genre, and categorizing music by mood, the *mel-frequency cepstral coefficient descriptors*, which are used for speech and speaker recognition and music modeling, and the *perceptual spread descriptor*, which represents the timbral width of sounds.

Local spatiotemporal feature descriptors, aggregates, and statistics capture aspects of both appearance and motion, and are used in video action recognition. *SIFT (Scale-invariant feature transform)*, as its name suggests, is a scale-invariant feature descriptor [1], which is suitable for object recognition, robotic navigation, 3D modeling, gesture recognition, and video tracking. The *cuboid descriptor* is a spatiotemporal interest point detector, which finds local regions of interest in space and time (cuboids) to be used for behavior recognition [2]. The *histogram of oriented gradients (HOG)* describes the number of occurrences of gradient orientation in localized portions of images and video frames [3], and is one of the most powerful feature statistics. HOG-based appearance descriptors combined with various motion descriptors based on the histogram of optical flow (HOF) are suitable for human detection in videos [4]. *Motion boundary histograms (MBH)* represent local orientations of motion edges by emulating static image HOG descriptors. The *speeded up robust features (SURF)* feature descriptor is based on the sum of the Haar wavelet response around the point of interest [5]. SURF is suitable for locating and recognizing objects and people, reconstructing 3D scenes, extracting points of interest, and object tracking.

Note that video analysis performed in the pixel domain is computationally expensive, and video analysis in the compressed domain, which is computationally cheaper, is not always an option. Also, the video container, format, and codec determines the implementation of feature extraction algorithms, such as the frame-level analysis of a Motion JPEG in an AVI container is different from that of an H.265/HEVC in a Matroska container. Owing to the compression algorithm used in MPEG videos, not all frames can be used directly for concept mapping, as many of the frames in MPEG files and video streams rely on previous frames.

6.1.3 Machine Learning in Video Content Analysis

The low-level video features can be automatically extracted using well-established algorithms, such as Gabor filter banks (extract homogeneous texture descriptors) [6] and fast color quantization (extracts dominant colors) [7]. Video content analysis employs advanced algorithms, such as the Viola-Jones and Lienhart-Maydt object detection algorithms [8, 9], and the SIFT [10], SURF [11], and ORB [12] keypoint detection algorithms. The corresponding low-level descriptors and frame regions can be used as positive and negative examples in machine learning, such as with support vector machines (SVM) and Bayesian networks, for keyframe analysis, face recognition, and similar tasks. However, low-level feature descriptors alone are not sufficient for video scene understanding, as will be demonstrated in the next section.

6.1.4 The Semantic Gap

What makes video understanding particularly challenging is the *Semantic Gap*, i.e., the huge discrepancy between what computers can interpret using automatically extracted low-level features and what humans understand based on cognition, knowledge, and experience [13]. For example, training from a few hundred or few thousand clips provides a very limited recognition capability, which cannot compete with years or decades of life experience and learning typical to humans. Training provides information for particular viewing angles only for the represented 3D space, although scale-/rotation-invariant features in 2D space can be used for object tracking in videos. For video processing algorithms, occlusion poses a real challenge, while recognizing partially covered objects is often very easy for humans. There are very few methods for complex video event detection, while humans understand even nonlinear narratives, such as extensive flashbacks and flash-forwards. On top of these, if the noise-signal ratio falls below a threshold, algorithms perform poorly. For these reasons, video understanding is often infeasible even without time constraints, let alone in near-real time or real time.

Automatically extractable low-level features and their statistics convey no information about the actual visual content, while humans can understand visual contents even without colors. In fact, most *low-level feature descriptors* are inadequate for representing *multimedia semantics* (the meaning of multimedia contents). For example, detecting red as a dominant color of a video frame (or one of its regions) does not provide information about the meaning of the visual content, which might be anything that contains red, and it cannot be inferred that a sunset is depicted [14]. Similarly, motion vectors are crucial in the motion estimation employed by video compression algorithms, but they do not correspond to the meaning of video scenes, because they only tell that someone or something is moving in a particular direction, which can be virtually anything.

Mid-level feature descriptors represent perceptual intuitions and higher-level semantics derived from signal-level saliency. Examples include concept hierarchy, visual patterns, segmented image patches, tracked objects, bags of features, spatial pyramids, and named entities extracted from subtitles. They are suitable for constructing expressive semantic features for visual content understanding and classification.

In contrast to low-level descriptors, *high-level descriptors* are suitable for multimedia concept mapping; however, they heavily rely on human knowledge, experience, and judgment. As such, the most sophisticated high-level descriptors are typically produced manually, which is a process that is not always feasible, and might be biased by the opinion or belief of, and influenced by the background of, the persons who perform the annotation. Some of the overly generic or inappropriate tags might be eliminated via the increasingly popular collaborative semantic video annotation, enabling multiple users to annotate video resources and improve existing annotations [15].

To narrow the Semantic Gap, feature extraction and analysis can be complemented by formally described *background knowledge*. This enables video interpretation by generating potential explanations and choosing the most likely one based on the maxima of preference scores. The rough modeling of background knowledge can capture the multiplicity of possible interpretations of a scene, using common sense knowledge and/or terminological knowledge that formally define the typical appearance of depicted objects.

6.2 Semantic Enrichment of Audiovisual Contents

There are various approaches to narrow, if not bridge, the Semantic Gap in videos. Video semantics, in the form of formal descriptions of video contents, utilize formal concept and property definitions from ontologies. The semantic enrichment of videos enables task automation via high-level scene interpretation and intelligent tasks, such as video event detection [16], moving object detection and tracking [17], intelligent video surveillance [18], and real-time activity monitoring [19]. Various knowledge representation techniques can be used for the spatiotemporal annotation of videos, so that still and moving regions of interests can be identified and uniquely referenced, as you will see in the following sections.

6.2.1 Video Semantics

Keywords, tags, labels, categories, genre, rating, and age rating, are frequently used on video sharing portals, which can be useful for video classification, but not necessarily for video understanding. A textual description, such as the plot, helps humans understand the story the video tells, but it is less useful for software agents, although they might retrieve some keywords from it via natural language processing. Recent efforts from the signal processing and natural language processing communities attempted to employ deep learning to automatically generate a complete natural sentence for describing video contents, called *video captioning* [20], although the output is often limited, prone to errors, and overly generic.

The meaning of video contents can be described with rich video semantics only, for example context, possible interpretations, the depicted era, the filming location, whether the video is a depiction, what mood it sets, what are the associated emotions, what symbols it features, and the ultimate message of the video.

3D semantics represented in 2D videos provide hidden meaning [21], such as depth, perspective, camera angle, and material characteristics (e.g., reflectivity, shininess). The fusion of 3D and video properties provides a higher level of semantic enrichment than those approaches that rely on video properties alone [22].

Frame-level video semantics include depicted concepts, whose formal definitions and properties can be retrieved from *ontologies* that "specify content-specific agreements for sharing and reuse of formally represented human knowledge among computer software" [23]. It is important to differentiate objects and humans, because when humans are detected, the gender might be determined, the faces recognized, and human behavior predicted. However, frame-based representations of video scenes are quite limited, because they miss out audio features and video events that happen over time.

Audio semantics include the recognition of the voice of a particular person, the recognition of particular sounds (e.g., gunshot) [24], and events, mood, emotions, and concepts associated with music [25]. Motion semantics of video scenes include moving objects, their speed (and whether it is constant, accelerating or decelerating), direction, and motion trajectory of objects and/or persons in a scene, the interaction of objects when a moving object hits another one (whether the moving object goes through or breaks the stationary one, or stops). These semantics, together with rules associated with the depicted scene, can form video events.

When utilizing background knowledge from ontologies, the formal description of these video semantics is suitable for content-based video retrieval, scene interpretation, and video understanding [26]. Some examples for background knowledge include concept hierarchy, relationship between concepts, and rules that define a *knowledge domain* (field of interest or area of concern):

- The hierarchy of depicted concepts defined with logical formalisms enables specialization or generalization via subclass-superclass relationships, such as marsupial is a subclass of the animal class. If koala is defined as a marsupial, and a koala is detected in a video, not only can be stated that a koala is depicted, but also that a marsupial is depicted and that an animal is depicted, which would not be possible at all with pure machine learning techniques (without knowledge representation).
- Objects depicted in videos usually do not appear in isolation as they are often correlated to each other. A corpora of concepts frequently depicted together can be used to set likelihood values for the correctness of concept mapping, hence the co-occurrence (semantic relatedness) of objects in temporal annotations adds an additional layer of semantics to videos [27]. For example, a kangaroo is very

likely to be depicted with an acacia tree of the Australian Outback, but more than unlikely with a giant sequoia, which is native to the US.

- Links to related resources enables knowledge discovery, advanced information retrieval, displaying useful information in hypervideo applications during video playback, and providing relevant videos that, based on a user's interests, are potentially interesting to the user.
- Rules are suitable for annotating complex video events and provide very rich semantics about the knowledge domain related to the video content. For example, by formally defining the competition rules of soccer, the number of players, the layout of the pitch, and match events, it is possible to automatically generate subtitles for soccer videos [28].

In scene interpretation, asserted knowledge provided by ontologies can be complemented by a priori knowledge obtained via rule-based *reasoning* [29].

6.2.2 Spatiotemporal Video Annotation Using Formal Knowledge Representation

Textual descriptions of multimedia resources constitute *unstructured data*, which is human-readable only [30]. For example, if a sentence in a natural language makes a statement about the running time of a movie as plain text, software agents can only process the string as meaningless consecutive characters. If the same information is written as *semistructured data*, such as in XML, it becomes machine-readable, so that computers can extract different entities and properties from the text (e.g., the running time can be declared and retrieved as a positive integer). However, the meaning of the number is still not defined. By leveraging organized, *structured data*, the previous code can be made machine-interpretable. Structured knowledge representations are usually expressed in, or based on, the *Resource Description Framework (RDF)*,[5] which can describe machine-readable statements in the form of subject-predicate-object (resource-property-value) triples, called *RDF triples*, e.g., scene-depicts-car (see Definition 6.1).

Definition 6.1 (RDF Triple). Assume there are pairwise disjoint infinite sets of

1) International Resource Identifiers ($\mathbb{I}$), i.e., sets of strings of Unicode characters of the form `scheme:[//[user:password@]host[:port]][/]path [?query][#fragment]` used to identify a resource,
2) RDF literals ($\mathbb{L}$), which are either a) self-denoting plain literals $\mathbb{L}_P$ in the form `"<string>"` `(@<lang>)?`, where `<string>` is a string and `<lang>` is an optional language tag, or b) typed literals $\mathbb{L}_T$ of the form `"<string>"^^<datatype>`, where `<datatype>` is an IRI denoting a datatype according to a schema (e.g., XML Schema), and `<string>` is an element of the lexical space corresponding to the datatype, and

---

[5]https://www.w3.org/TR/rdf11-concepts/.

3) Blank nodes ($\mathbb{B}$), i.e., unique but anonymous resources that are neither IRIs nor RDF literals.

A triple $(s, o, p) \in (\mathbb{I} \cup \mathbb{B}) \times \mathbb{I} \times (\mathbb{I} \cup \mathbb{L} \cup \mathbb{B})$ is called an *RDF triple* (or RDF statement), where s is the subject, p is the predicate, and o is the object.

The corresponding classes, properties, and relationships are typically defined in controlled vocabularies (see Definition 6.2) or ontologies written in the first or second version of the *Web Ontology Language* (OWL or OWL 2), which are formally grounded in *description logics (DL)*.[6]

Definition 6.2 (Controlled Vocabulary). A controlled vocabulary is a triple $V = (N_C, N_R, N_I)$ of countably infinite sets of IRI symbols denoting atomic concepts (concept names or classes) (N_C), atomic roles (role names, properties, or predicates) (N_R), and individual names (objects) (N_I), respectively, where N_C, N_R, and N_I are pairwise disjoint sets.

For example, a wildlife vocabulary may have classes such as Mammal and Bird that form set N_C, properties such as scientificName and isEndangered as well as relations such as closeRelativeOf and preysOn that form set N_R, and individuals such as PLATYPUS and EMU that form set N_I.[7]

These formal knowledge representation languages support different sets of mathematical constructors to achieve a favorable trade-off between expressivity and computational complexity (which depends on the intended application). For example, the $\mathcal{ALC}$ description logic supports atomic negation, concept intersection, universal restrictions, limited existential quantification, and complex class expressions using a combination of operators, such as subclass relationships, equivalence, conjunction, disjunction, negation, property restrictions, tautology, and contradiction. $\mathcal{ALC}$ extended with transitive roles is called $\mathcal{S}$. If all the previous constructors are extended with $\mathcal{H}$ (role hierarchy), $\mathcal{O}$ (enumerated concept individuals), $\mathcal{I}$ (inverse roles), and $\mathcal{N}$ (unqualified cardinality restrictions), the description logic is called $\mathcal{SHOIN}$, which roughly corresponds to OWL DL. Adding $\mathcal{R}$ (complex role inclusion, reflexivity and irreflexivity, and role disjointness) to the above and replacing $\mathcal{N}$ with $\mathcal{Q}$ (qualified cardinality restrictions) yields to $\mathcal{SROIQ}$, which is the description logic behind OWL 2 DL (see Definition 6.3). Those description logics that support datatypes, datatype properties, and data values also feature a trailing $^{(\mathcal{D})}$ superscript in their names.

---

[6]OWL classes and properties correspond to description logic concepts and roles. Individuals are called the same way in both the OWL and the description logic terminology.

[7]The description logic concepts and roles do not follow the general capitalization rules of English grammar; instead, they purposefully capitalize each word to make them easier to read. In concrete examples for concept names, the first letter of the identifier and the first letter of each subsequent concatenated word are capitalized (PascalCase), for role names, the first letter of the identifier is lowercase and the first letter of each subsequent concatenated word is capitalized (camelCase), and individual names are written in ALL CAPS.

Definition 6.3 ($\mathcal{SROIQ}$ Ontology).[8] A $\mathcal{SROIQ}$ *ontology* is a set of role expressions **R** over a signature defined as $\mathbf{R} :: = U|N_R|N_R^-$, where U represents the universal role, N_R is a set of roles, and N_R^- is a set of negated role assertions. The concept expressions of a $\mathcal{SROIQ}$ ontology are defined as the set $\mathbf{C} :: = N_C|(C \sqcap D)|(C \sqcup D)|\neg C|\top|\bot|\exists R.C|\forall R.C| \geqslant nR.C| \leqslant nR.C|\exists R.Self|\{N_I\}$, where n is a non-negative integer, C and D represent concepts, and R represents roles. Based on these sets, $\mathcal{SROIQ}$ axioms can be defined as general concept inclusions (GCIs) of the form $C \sqsubseteq D$ and $C \equiv D$ for concepts C and D (terminological knowledge, TBox), individual assertions of the form $C(N_I)$, $R(N_{I_1}, N_{I_2})$, $N_{I_1} \approx N_{I_2}$, or $N_{I_1} \not\approx N_{I_2}$ (assertional knowledge, ABox), and role assertions of the form $R \sqsubseteq S, R \equiv S, R_1 \circ \ldots \circ R_n \sqsubseteq S, Asymmetric(R), Reflexive(R), Irreflexive(R), Disjoint(R, S)$[9] for roles R, R_i, and S (role box, RBox) [31].

For example, a car ontology may contain, among others, concepts such as Vehicle, Car, LuxuryCar, and Fuel, individuals such as Chrysler300C, LanciaThema, Petrol, Diesel, and Biodiesel, and roles such as rebadgeOf, basedOn, poweredBy, and hasEngine. These can be defined as GCIs such as Car $\sqsubseteq$ Vehicle and Car $\equiv$ Automobile, individual assertions such as Car(CHRYSLER300C), Car(LANCIATHEMA), Fuel(PETROL), Fuel(DIESEL), Fuel(BIODIESEL), type(CHRYSLER300C, LuxuryCar),[10] rebadgeOf(LANCIATHEMA, CHRYSLER300C), poweredBy (CHRYSLER300C, PETROL $\sqcup$ DIESEL), GASOLINE $\approx$ PETROL, and DIESEL $\not\approx$ BIODIESEL, and role assertions such as rebadgeOf $\sqsubseteq$ basedOn and poweredBy $\equiv$ usesFuel.

The meaning of description logic concepts and roles is defined by their model-theoretic semantics, which are based on *interpretations*. In $\mathcal{SROIQ}$, interpretation $\mathcal{I}$ consists of a set $\Delta^\mathcal{I}$ (the domain of $\mathcal{I}$) and an interpretation function $\cdot^\mathcal{I}$, which maps each atomic concept A to a set $A^\mathcal{I} \subseteq \Delta^\mathcal{I}$, each atomic role R to a binary relation $R^\mathcal{I} \subseteq \Delta^\mathcal{I} \times \Delta^\mathcal{I}$, and each individual name a to an element $a^\mathcal{I} \in \Delta^\mathcal{I}$.

The formal definition of concepts and roles used in description logic formalisms are defined in controlled vocabularies and ontologies, which will be described in the following sections.

6.2.3 Vocabularies and Ontologies

Vocabularies, taxonomies, thesauri, and simple ontologies are usually defined in *RDF Schema (RDFS)*,[11] an extension of RDF specially designed for defining taxonomical

---

[8]The formal definition of an ontology depends on its logical underpinning, but the most expressive OWL 2 ontologies defined here are supersets of all the ontologies that utilize less expressive formalisms. Most OWL 2 ontologies do not exploit all the available mathematical constructors of the underlying logical underpinning.

[9]Often abbreviated with the first three letters as *Asy*(R), *Ref*(R), *Irr*(R), and *Dis*(R, S).

[10]The `type` (`isA`) relationship is typically reused from the RDF vocabulary (`rdf:type`).

[11]https://www.w3.org/TR/rdf-schema/.

structures and concept relationships, while complex ontologies are defined in the fully-featured ontology language OWL.

The declaration of a video clip depicting a person in a machine-readable format, for example, requires the formal definition of video clips and their features to be retrieved from a vocabulary or ontology, such as the Clip vocabulary from Schema.org,[12] which is suitable for declaring the director, file format, language, encoding, and other properties of video clips (schema:Clip). The definition of the "depicts" relationship can be found in the *Video Ontology* (*VidOnt*)[13] (video:depicts), a state-of-the-art core reference ontology for video, which integrates viewpoints of de facto standard and standard video-related upper ontologies and domain ontologies with important concepts and roles, constituting the most expressive video ontology to date. The definition of "Person" can be used from schema:Person, which defines typical properties of a person, including, but not limited to, name, gender, birthdate, and nationality.[14]

The vocabularies of core audio and video metadata standards have originally been created in XML or XML Schema (XSD),[15, 16, 17] which made them machine-readable, but not machine-interpretable. *Semantic Web* standards, such as RDF, RDFS, and OWL, can overcome this limitation [32], which resulted in several attempts for the RDFS or OWL mapping of ID3,[18] Dublin Core,[19] TV-Anytime,[20] MPEG-7,[21] or a combination of these [33]. Hunter's MPEG-7 ontology was the first of its kind; it modeled the core parts of MPEG-7 in OWL Full, complemented by DAML+OIL constructs [34]. Inspired by this mapping, Tsinaraki et al. created another MPEG-7 ontology, but with full coverage of the MPEG-7 Multimedia Description Scheme (MDS) [35]. *Rhizomik* (MPEG-7Ontos),[22] the first complete MPEG-7 ontology, was generated using a transparent mapping from XML to RDF combined with mapping XSD to OWL [36]. The *Visual Descriptor Ontology (VDO)* was an OWL DL ontology, which covered the visual components of MPEG-7 [37]. The *Multimedia Structure Ontology (MSO)* defined basic multimedia concepts from the MPEG-7 MDS [38]. Oberle et al. created an ontological framework to formally model the MPEG-7 descriptors and export them to OWL [39]. The *Multimedia Content Ontology (MCO)* and the *Multimedia Descriptors Ontology (MDO)* cover the MPEG-7

---

[12]http://schema.org/Clip.

[13]http://videoontology.org.

[14]It is a common practice to abbreviate terms using the *namespace mechanism*, which uses a prefix instead of full (and often symbolic) URIs. For example, schema: abbreviates http://schema.org/, therefore schema:Clip stands for http://schema.org/Clip.

[15]http://standards.iso.org/ittf/PubliclyAvailableStandards/MPEG-7_schema_files/.

[16]http://purl.org/NET/mco-core, http://purl.org/NET/mco-ipre.

[17]http://webapp.etsi.org/workprogram/Report_WorkItem.asp?WKI_ID=39864.

[18]http://www.semanticdesktop.org/ontologies/2007/05/10/nid3/.

[19]http://dublincore.org/2012/06/14/dcterms.rdf.

[20]http://rhizomik.net/ontologies/2005/03/TVAnytimeContent.owl.

[21]http://mpeg7.org.

[22]http://rhizomik.net/ontologies/2005/03/Mpeg7-2001.owl.

MDS structural descriptors, and the visual and audio parts of MPEG-7 [40]. The *Core Ontology for Multimedia* (*COMM*) mapped selected parts of the MPEG-7 vocabulary to OWL [41]. The X3D-aligned *3D Modeling Ontology*[23] can be used for describing characters and objects of computer animations, including computer generated imagery (CGI) and computer-aided design (CAD), as well as virtual, augmented, and mixed reality videos [42].

Wordnet[24] and *OpenCyc*[25] are two well-established upper ontologies that can be used for describing a variety of concepts, including those depicted in videos. Alternatively, ontologies specially designed for this purpose, such as the *Large-Scale Concept Ontology for Multimedia* (*LSCOM*) [43] or the ontology of Zha et al. [44] can also be used.

The spatiotemporal annotation of video events may employ spatial ontologies, such as RCC-8 calculus-based ontologies [45], temporal ontologies, such as the *SWRL Temporal Ontology*,[26] and fuzzy ontologies, such as the *Video Semantic Content Model* (*VISCOM*) [46], which are formally grounded in not only general, but also in spatial, temporal, and fuzzy description logics [47].[27]

Schema.org provides de facto standard definitions for a variety of knowledge domains, which also includes coverage for concepts and properties that frequently appear in multimedia contents. For example, audio resources can be described with `schema:bitrate`, `schema:encodingFormat`, and `schema:duration`. Similarly, videos can be described using `schema:video` and `schema:VideoObject`. Seasons, film series, episodes of series, and movies can be annotated with `schema:CreativeWorkSeason`, `schema:Movie Series`, `schema:Episode`, and `schema:Movie`. Genres can be defined using `schema:genre`.

In addition, several OWL ontologies have been created for de facto standards and many without standards alignment [48]. W3C's *Ontology for Media Resources*[28] provides a core vocabulary with standards alignment to be used in online media resource descriptions. The *Multimedia Metadata Ontology* (*M3O*)[29] was designed to integrate the core aspects of multimedia metadata [49]. The *Linked Movie Database*[30] was designed for describing common concepts and properties of Hollywood movies, such as actor, director, etc. The *STIMONT* ontology can describe emotions associated with video scenes [50].

The terms of these ontologies, when serialized in RDFa, HTML5 Microdata, or JSON-LD, can be indexed by all major search engines, including Google, Yahoo!, and Bing [51].

---

[23]http://3dontology.org.

[24]http://wordnet-rdf.princeton.edu/ontology.

[25]https://sourceforge.net/projects/texai/files/open-cyc-rdf/1.1/.

[26]http://swrl.stanford.edu/ontologies/built-ins/3.3/temporal.owl.

[27]Not all of these formalisms can be implemented in OWL 2, leading to proprietary extensions.

[28]http://www.w3.org/TR/mediaont-10/.

[29]http://m3o.semantic-multimedia.org/ontology/2009/09/16/annotation.owl.

[30]http://www.linkedmdb.org.

6.2.4 Semantic Enrichment of Videos with Linked Data

Records of isolated video databases, particularly when locked down and using proprietary data formats, are inefficient in data access, sharing, and reuse.

To enable semantic queries across diverse resources, structured data is often published according to best practices (*Linked Data*) [52]. The four principles of Linked Data are the following[31]:

1. Uniform Resource Identifiers (URIs), i.e., strings of ASCII characters of the form `scheme:[//[user:password@]host[:port]][/]path [?query][#fragment]`, should be used to represent real-world concepts and entities.
2. The URIs of RDF resources should be HTTP URIs, so that the resource names can be found on the Internet.
3. The resource URIs should provide useful information using Semantic Web standards, such as RDF.
4. Links to other URIs should be included, enabling users and software agents discover related information.

Creating links between the structured datasets of the Semantic Web is called *interlinking*, which makes isolated datasets part of the LOD Cloud,[32] in which all resources are linked to one another. These links enable semantic agents to navigate between data sources (traverse RDF graphs) and discover related resources. The most common predicates used for interlinking are `owl:sameAs` and `rdfs:seeAlso`, but any predicate can be used. In contrast to hyperlinks between web pages, LOD links utilize typed RDF links between resources.

Linked Data with an explicitly stated open license is called *Linked Open Data* (*LOD*). A meaningful collection of RDF triples that complies with Linked Data principles and is published with an open license is called an *LOD dataset*. The LOD-based semantic enrichment of videos is employed by video repositories, hypervideo applications, and video streaming portals, such as YouTube [53].

6.2.5 Spatiotemporal Annotation in Action

To demonstrate spatiotemporal annotation in action, consider the argument scene from the movie "The Sound of Music" with Maria and Captain von Trapp, portrayed by Julie Andrews and Christopher Plummer (20th Century Fox, 1965). The aim is to identify the video scene with temporal data, annotate the region of interest depicting Maria as a still region and the region of interest depicting the captain as a moving region with spatiotemporal segmentation, and describe the movie scene, the two movie characters, and the actors who played in the corresponding roles. By

---

[31] https://www.w3.org/DesignIssues/LinkedData.html.

[32] http://lod-cloud.net.

Fig. 6.1 Spatial annotation of regions of interest with the top left corner coordinates, width, and height of the minimum bounding boxes. Movie scene by 20th Century Fox

using Media Fragment URI 1.0 identifiers,[33] the spatiotemporal segmentation can be done as follows. The positions of the selected shots are specified in Normal Play Time format according to RFC 2326.[34] The movie characters are represented by their minimum bounding boxes, as shown in Fig. 6.1.

Using a description logic formalism, this video scene can be represented as shown in Listing 6.1.

Listing 6.1 Spatiotemporal description of a video scene with DL formalism

```
Movie(THESOUNDOFMUSIC)
filmAdaptationOf(THESOUNDOFMUSIC,
THESTORYOFTHETRAPPFAMILYSINGERS)
Scene ⊑ VideoSegment
Scene(ARGUMENTSCENE)
sceneFrom(ARGUMENTSCENE, THESOUNDOFMUSIC)
hasStartTime(ARGUMENTSCENE, 01:12:48)
duration(ARGUMENTSCENE, 00:01:54)
hasFinishTime(ARGUMENTSCENE, 01:14:42)
depicts(ARGUMENTSCENE, Argument)
MovieCharacter(MARIA)
portrayedBy(MARIA, JULIEANDREWS)
MovieCharacter(CAPTAINVONTRAPP)
portrayedBy(CAPTAINVONTRAPP, CHRISTOPHERPLUMMER)
partOf(ARGUMENTROI1, ARGUMENTSCENE)
partOf(ARGUMENTROI2, ARGUMENTSCENE)
StillRegion(ARGUMENTROI1)
```

---

[33]https://www.w3.org/TR/media-frags/.

[34]https://www.ietf.org/rfc/rfc2326.txt.

MovingRegion(ARGUMENTROI2)
depicts(ARGUMENTROI1, MARIA)
depicts(ARGUMENTROI2, CAPTAINVONTRAPP)

This formal description can be written in any RDF serialization, such as
RDF/XML, Turtle, Notation3 (N3), N-Triples, N-Quads, and any compatible
lightweight annotation, such as RDFa, HTML5 Microdata, and JSON-LD. Listing
6.2 shows the Turtle serialization of the above example.

Listing 6.2 Spatiotemporal description of a video scene in Turtle

```
@prefix dbpedia: <http://dbpedia.org/resource/> .
@prefix mpeg-7: <http://purl.org/ontology/mpeg7/> .
@prefix temporal: <http://swrl.stanford.edu/ontologies/built-
ins/3.3/temporal.owl> .
@prefix schema: <http://schema.org/> .
@prefix video: <http://purl.org/ontology/video/> .
@prefix xsd: <http://www.w3.org/2001/XMLSchema#> .
dbpedia:The_Sound_of_Music_(film) a video:movie ;
 video:filmAdaptationOf
 dbpedia:The_Story_of_the_Trapp_Family_Singers .
<http://example.com/soundofmusic.mp4> a mpeg-7:Video ,
 schema:Movie .
<http://example.com/soundofmusic.mp4#t=1:12:48,1:14:42> a
 video:Scene ; video:temporalSegmentOf
 <http://example.com/soundofmusic.mp4> ;
 video:sceneFrom dbpedia:The_Sound_of_Music_(film) ;
 temporal:hasStartTime "01:12:48"^^xsd:time ;
 temporal:duration "PT01M54S"^^xsd:duration ;
 temporal:hasFinishTime "01:14:42"^^xsd:time ;
 video:depicts dbpedia:argument .
dbpedia:Maria_von_Trapp a vidont:MovieCharacter ;
 video:portrayedBy dbpedia:Julie_Andrews .
dbpedia:Georg_von_Trapp a vidont:MovieCharacter ;
 video:portrayedBy dbpedia:Christopher_Plummer .
<http://example.com/soundofmusic.mp4#t=1:12:49,1:12:53&xywh=
1074,293,302,584> a mpeg-7:MovingRegion ;
 video:spatioTemporalSegmentOf
 <http://example.com/soundofmusic.mp4> ;
 video:depicts dbpedia:Georg_von_Trapp .
<http://example.com/soundofmusic.mp4#t=1:12:49,1:12:53&xywh=
164,51,454,827> a mpeg-7:StillRegion ;
 video:spatioTemporalSegmentOf
 <http://example.com/soundofmusic.mp4> ;
 video:depicts dbpedia:Maria_von_Trapp .
```

The formal definition of the terms used in the video scene description above are retrieved from MPEG-7, VidOnt, the SWRL Temporal Ontology, DBpedia, and Schema.org, by declaring their namespaces (see the lines starting with @prefix) and using the corresponding prefixes in the RDF triples. In Turtle, a is a shorthand notation for the rdf:type predicate, which expresses an "is a" relationship. The above example uses another shorthand notation as well, namely that a series of RDF triples sharing the same subject can be abbreviated by stating the subject once, and then each predicate-object pair separated using a semicolon. Temporal segments are identified for the soundofmusic.mp4 video file by stating the starting and ending time separated by a comma, preceded by #t=, which is, like any URI in Turtle, delimited by < and >. The spatiotemporal segment for the region of interest extends this by the top left coordinates and dimensions of the minimum bounding box of the region separated by commas and preceded by &xywh= in the URI (a spatial segment of the temporal segment). Some of these annotations can be generated using semantic video annotation tools, although the RDF output varies greatly due to proprietary ontology implementations [54].

In contrast to the tree structure of XML documents, RDF-based knowledge representations can be visualized as graphs. RDF graphs are directed, labeled graphs in which the nodes are the resources and values, and the arrows assign the predicates (see Fig. 6.2).

Because the RDF graphs that share the same resource identifiers naturally merge together, interlinking LOD concepts and individuals (e.g., dbpedia:argument, dbpedia:Maria_von_Trapp) makes the above graph part of the LOD Cloud. By traversing the interconnected graphs of the LOD Cloud, or by directly querying them, it is possible to find useful, relevant machine-interpretable information related to the depicted concepts, such as the memoir the story of the film is based on, further adaptations of the book, filming locations, the birthday of the staff members, and so on. Upon these, new information can also be inferred, such as the age of the actors and actresses at the time of filming can be calculated.

6.3 Ontology-Based Video Scene Interpretation

A common approach to ontology-based video understanding is the automated shot annotation with semantic labels using pretrained classifiers. However, frame-level object mapping alone is often insufficient to understand the visual content. To address this limitation, events can be used to provide additional information to interpret a scene. Such information includes object positions, object transitions over time, and the relationship between objects and high-level concepts. Formalizing complex events by combining primitive events is most efficient in well-defined domains with constrained actions and environment (e.g., soccer videos). While this chapter focuses on knowledge-based techniques, there are other high-level video recognition approaches in unconstrained videos, such as the ones that utilize bag of features, kernel classifiers, and multimodal fusion [55].

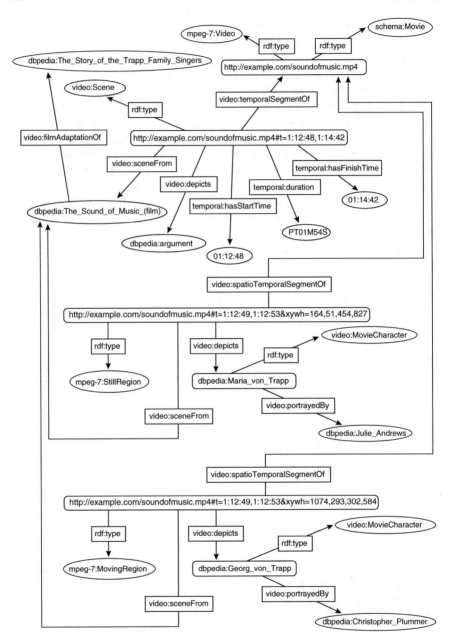

Fig. 6.2 Graph visualization of the RDF triples of Listing 6.2

Many knowledge-based high-level scene interpretation tasks are performed by deductive reasoning over the video contents, during which new statements are inferred based on explicit ontology statements, to recognize situations and temporal events based on human knowledge formally described as ontology concepts, roles, individuals, and rules. By representing fuzzy relationships between the context and depicted concepts of video contents, abductive reasoning can also be performed [56].

6.3.1 Video Event Recognition via Reasoning Over Temporal DL Axioms

Video event recognition often requires higher expressivity than what is available in general description logic formalisms, such as the one demonstrated in the previous section. Complex video events can be formally described and automatically recognized via reasoning by using temporal description logics, such as $\mathcal{TL}$-$\mathcal{F}$, which can be briefly defined as follows. $\mathcal{TL}$-$\mathcal{F}$ is composed of the temporal language $\mathcal{TL}$, which expresses interval temporal networks, and a non-temporal feature description logic $\mathcal{F}$. The $\mathcal{TL}$ part of $\mathcal{TL}$-$\mathcal{F}$ can contain non-temporal concepts (E), conjunction ($C \sqcap D$), qualifiers ($C@X$),[35] substitutive qualifiers ($C[Y]@X$), temporal constraints (($X(R)Y$), ($X(R)\sharp$), and ($\sharp(R)Y$)), existential quantifiers ($\diamond(\bar{X})\bar{T}_C.C$),[36] disjunction ($R, S$), temporal variables ($x, y, z, \ldots$), and Allen's relations: b (*before*), m (*meets*), d (*during*), o (*overlaps*), s (*starts*), f (*finishes*), = (*equal*), a (*after*), mi (*met by*), di (*contains*), oi (*overlapped by*), si (*started by*), and fi (*finished by*) [57]. The $\mathcal{F}$ part of $\mathcal{TL} - \mathcal{F}$ can contain atomic concepts (A), tautology ($\top$), conjunctions ($E \sqcap F$), agreements ($p \downarrow q$), selections ($p : E$), atomic features (f), atomic parametric features ($\bigstar g$),[37] and paths ($p \circ q$).

Using $\mathcal{TL}$-$\mathcal{F}$ formalisms, a goal in soccer videos, for example, can be described by the following sequence: goalpost–cheers–closeup–audience–slow motion replay [58] (see Listing 6.3).

Listing 6.3 Formal description of a video event using the temporal description logic $\mathcal{TL}$-$\mathcal{F}$

GOAL $= \diamond(d_{goal}, d_{whistle}, d_{cheers}, d_{caption}, d_{goalpost}, d_{closeup}, d_{audience}, d_{replay})$ $(d_{goal}$ f $d_{goalpost})(d_{whistle}$ d $d_{goalpost})(d_{goalpost}$ o $d_{cheers})(d_{caption}$ e $d_{closeup})(d_{cheers}$ e $d_{closeup})$ $(D_{goalpost}$ m $d_{closeup})(d_{closeup}$ m $d_{audience})(d_{audience}$ m $d_{MSR})$.(GOAL @ d_{goal} $\sqcap$ WHISTLE @ $d_{whistle}$ $\sqcap$ CHEERS @ d_{cheers} $\sqcap$ CAPTION @ $d_{caption}$ $\sqcap$ GOALPOST @ $d_{GOALPOST}$ $\sqcap$ CLOSEUP @ $d_{CLOSEUP}$ $\sqcap$ AUDIENCE @ $d_{AUDIENCE}$ $\sqcap$ REPLAY @ $d_{replay})$

---

[35]$C@X$ enables the evaluation of concept C at an interval X different from the current one by temporally qualifying it at X.

[36]The temporal existential quantifier $\diamond$ introduces interval variables related to each other and to variable $\sharp$ (now, a special temporal variable which serves as a reference) according to a set of temporal constraints.

[37]$\bigstar$ distinguishes parametric and non-parametric features and is not an operator.

where $\diamond$ is the temporal existential quantifier for introducing the temporal intervals, @ is a bindable variable, and d_{goal}, $d_{whistle}$, d_{cheers}, $d_{caption}$, $d_{goalpost}$, $d_{closeup}$, $d_{audience}$, and d_{replay} represent the temporal intervals of the corresponding objects and sequences. After detecting the objects and sequences in a soccer video, they can be described in the form $\diamond x(\ ).C@x$, where C is the individual of the object or sequence, x is the temporal interval of C, and ( ) denotes those individuals that do not have temporal relationships. Assume a set of sequence individuals $\{S_0, S_1, \ldots, S_{n-1}, S_n\}$ from the detection results of a soccer video, in which each element S_i can be represented in the form $S_i = \diamond x_i(\ ).S_i@x_i$. The definition of $\{S_0, S_1, \ldots, S_{n-1}, S_n\}$ includes a latent temporal constraint, $x_i m x_{i+1}$, $i = 0, 1, \ldots, n-1$, which denotes two consecutive sequences in $\{S_0, S_1, \ldots, S_{n-1}, S_n\}$ that are consecutive in the temporal axis of the video. Further assume a set of object individuals $\{O_0, O_1, \ldots, O_{m-1}, O_m\}$ from the detection results of a soccer video, in which each element O_i can be represented in the form $O_i = \diamond y_i(\ ).O_i@y_i$. Based on the above representation, reasoning can be performed over soccer videos to recognize goals as follows. Firstly, those subsets of $\{S_0, S_1, \ldots, S_{n-1}, S_n\}$ are selected that are composed of the consecutive individuals GOALPOST » CLOSEUP » AUDIENCE » REPLAY. These subsets are all goal candidates calculated as $E_{Ck} = \{$GOALPOST$_k$, CLOSEUP$_{k+1}$, AUDIENCE$_{k+2}$, REPLAY$_{k+3}\}$, where k is the index of the current view of the current candidate event in $\{S_0, S_1, \ldots, S_{n-1}, S_n\}$. Secondly, all the goal objects O_{goal}, $O_{whistle}$, O_{cheers}, $O_{caption}$ have to be found in $\{O_0, O_1, \ldots, O_{m-1}, O_m\}$ for each candidate event E_{Ck}, which have a corresponding temporal interval (y_{goal}, $y_{whistle}$, y_{cheers}, $y_{caption}$), and satisfy the corresponding temporal constraints, i.e., $y_{goal} f$ GOALPOST$_k$, $y_{whistle} d$ GOALPOST$_k$, GOALPOST$_k$ $o y_{cheers}$, $y_{caption} e$ CLOSEUP$_{k+1}$, $y_{cheers} e$ CLOSEUP$_{k+1}$. If all of these objects exist, E_{Ck} can be considered a goal.

6.3.2   Video Event Recognition Using SWRL Rules

Description logic-based semantic video annotations can be complemented by rule-based representations, such as SWRL rules, to recognize video events and improve the integrity and correctness of the interpretation. In news videos, for example, SWRL rules enable the formal definition of the appearance of anchorpersons (see Listing 6.4).

Listing 6.4 Formal definition of a video event with SWRL rules [59]

```
Person(?p1) ^ hasValidPeriod(?p1, ?Vtip1) ^
hasValidPeriod(?p1, ?Vtip2) ^ hasValidPeriod(?p1, ?Vtip3) ^
differentFrom(?Vtip1, ?Vtip2) ^ differentFrom(?Vtip2, ?Vtip3)
^ differentFrom(?Vtip1, ?Vtip3) ^
temporal:hasFinishTime(?Vtip1, ?FTp1) ^
temporal:hasStartTime(?Vtip2, ?STp2) ^
temporal:hasFinishTime(?Vtip2, ?FTp2) ^
```

```
temporal:hasStartTime(?Vtip3, ?STp3) ^
temporal:duration(?dp1dp2,?FTp1, ?STp2, temporal:Seconds) ^
temporal:duration(?dp2dp3,?FTp2,?STp3, temporal:Seconds) ^
swrlb:greaterThan(?dp1dp2, 120) ^ swrlb:greaterThan(?dp2dp3,
120) ^ StudioSetting(?s1) ^ hasValidPeriod(?s1, ?Vtis1) ^
hasValidPeriod(?s1, ?Vtis2) ^ hasValidPeriod(?s1, ?Vtis3) ^
differentFrom(?Vtis1, ?Vtis2) ^ differentFrom(?Vtis2, ?Vtis3)
^ differentFrom(?Vtis1, ?Vtis3) ^ temporal:equals(?Vtis1,
?Vtip1,
temporal:Seconds) ^ temporal:equals(?Vtis2, ?Vtip2,
temporal:Seconds) ^ temporal:equals(?Vtis3, ?Vtip3,
temporal:Seconds) -> ? Anchor(?p1)
```

In this case, if person p1 appears in three different time intervals of a news broadcast (Vtip1, Vtip2, Vtip3), the occurrences have a temporal distance greater than the defined threshold of 120 seconds, and the occurrence of the StudioSetting instance coincides with the previous intervals (Vtis1, Vtis2, Vtis3), person p1 is considered an anchorperson.

The semantically enriched representation can be used by automated mechanisms to recognize the same type of video scenes in different video resources. Moreover, reasoners can use such machine-interpretable descriptions to automatically infer new statements, thereby achieving knowledge discovery.

Note, however, that the chosen formalism heavily affects computational complexity, and similar to many temporal description logics, SWRL rules might break decidability. Hence, such formalisms should be used for video annotations only when the expressivity of general description logics is insufficient.

6.3.3 Handling the Uncertainty of Concept Depiction with Fuzzy Axioms

Fuzzy description logics provide inference support for vague information, which can be utilized in video frame interpretation tasks, such as object recognition [60]. For example, suppose the background knowledge of TBox and RBox axioms of Listing 6.5 to be used for reasoning-based video frame interpretation.

Listing 6.5 Formally described background knowledge

$\mathcal{T} \equiv \{\langle$ Crown $\sqsubseteq$ color(green) $\sqcap$ texture(patchy)$\rangle$, $\langle$Trunk $\sqsubseteq$ color(brown) $\sqcap$ texture(rough)$\rangle$, $\langle$Tree $\equiv \exists$hasPart.(Trunk $\sqcap \exists$isBelow.Crown)$\rangle\}$

$\mathcal{R} \equiv \{$Trans(hasPart)$\}$

Assume a frame depicting a group of trees segmented, and a set of values produced for each region based on their color and texture. In fuzzy description logics, these values can be described with fuzzy ABox assertions as shown in Listing 6.6.

Listing 6.6 Fuzzy Abox axioms

$\mathcal{A} \equiv \{\langle color(o1, green)\rangle \geqslant 0.85\rangle, \langle texture(o1, patchy) \geqslant 0.7\rangle, \langle color(o2, brown) \geqslant 1.0\rangle, \langle texture(o2, rough) \geqslant 0.9\rangle, \langle isAbove(o1, o2) \geqslant 0.9\rangle, \langle hasPart(o3, o2) \geqslant 0.8\rangle\}$

A fuzzy interpretation $\mathcal{I}$ is a model with reference to the TBox if it holds the statements described in Listing 6.7.

Listing 6.7 Fuzzy interpretation

$\mathrm{Crown}^{\mathcal{I}}(o_1^{\mathcal{I}}) = t(\mathrm{color}^{\mathcal{I}}(o_1^{\mathcal{I}}, \mathrm{green}^{\mathcal{I}}), \mathrm{texture}^{\mathcal{I}}(o_1^{\mathcal{I}}, \mathrm{patchy}^{\mathcal{I}})) = t(0.85, 0.7)$

$\mathrm{Trunk}^{\mathcal{I}}(o_2^{\mathcal{I}}) = t(\mathrm{color}^{\mathcal{I}}(o_2^{\mathcal{I}}, \mathrm{brown}^{\mathcal{I}}), \mathrm{texture}^{\mathcal{I}}(o_2^{\mathcal{I}}, \mathrm{rough}^{\mathcal{I}})) = t(1.0, 0.9)$

$\mathrm{Tree}^{\mathcal{I}}(o_3^{\mathcal{I}}) = \sup_{b}\{t(\mathrm{hasPart}^{\mathcal{I}}(o_3^{\mathcal{I}}, b), (\mathrm{Trunk} \sqcap \exists \mathrm{isBelow.Crown})^{\mathcal{I}}(b))\}$

$= \sup_{b}\{t(\mathrm{hasPart}^{\mathcal{I}}(o_3^{\mathcal{I}}, b), t(\mathrm{Trunk}^{\mathcal{I}}(b), \sup_{c}\{t(\mathrm{isBelow}^{\mathcal{I}}(b, c), \mathrm{Crown}^{\mathcal{I}}(c))\}))\}$

$\geqslant t(\mathrm{hasPart}^{\mathcal{I}}(o_3^{\mathcal{I}}, o_2^{\mathcal{I}}), t(\mathrm{Trunk}^{\mathcal{I}}(o_2^{\mathcal{I}}), t((\mathrm{isAbove}^-)^{\mathcal{I}}(o_2^{\mathcal{I}}, o_1^{\mathcal{I}}), \mathrm{Crown}^{\mathcal{I}}(o_1^{\mathcal{I}})))) \geqslant$

$t(0.8, t(t(1.0, 0.9), t(0.9, t(0.85, 0.7))))$

where t represents the fuzzy intersection performed by a function of the form $t :$ $[0, 1] \times [0, 1] \rightarrow [0, 1]$, called the *t-norm operation*, which must be commutative, i.e., $t(a, b) = t(b, a)$, monotonically increasing, i.e., for $a \leqslant c$ and $b \leqslant d$, $t(a, b) \leqslant t(c, d)$, and associative, i.e., $t(a, t(b, c)) = t(t(a, b), c)$, with 1 being an identity element, i.e., $t(a,1) = a$. Depending on the t-norm used, different values can be inferred for $o_3^{\mathcal{I}}$ being a tree. For example, in case of the Łukasiewicz t-norm, i.e., $\mathsf{T}_{\mathrm{Luk}}(a, b) = \max\{0, a+b-1\}$, $\mathrm{Tree}^{\mathcal{I}}(o_3^{\mathcal{I}}) \geqslant 0.15$, the product t-norm, i.e., $\mathsf{T}_{\mathrm{prod}}(a, b) = a \cdot b$, gives $\mathrm{Tree}^{\mathcal{I}}(o_3^{\mathcal{I}}) \geqslant 0.385$, while the minimum t-norm (Gödel t-norm), i.e., $\mathsf{T}_{\mathrm{min}}(a, b) = \min\{a, b\}$, yields to $\mathrm{Tree}^{\mathcal{I}}(o_3^{\mathcal{I}}) \geqslant 0.7$.

6.4 Utilizing Video Semantics: From Intelligent Video Indexing to Hypervideo Applications

Structured video annotation enables efficient data sharing and reuse, and task automation. The machine-interpretable description of events and spatiotemporal annotation of video scenes can be used to retrieve video scenes that are related to, or visually similar to, a reference scene. Rich semantics can also be displayed by web search engines and hypervideo applications, and used for additional services, such as automated video summaries, automated video classification, and video accessibility via providing video descriptions to screen readers.

6.4.1 Content-Based Video Indexing and Retrieval

Concept relationships are valuable knowledge resources that can enhance the effectiveness of video retrieval even for ambiguous queries [61]. RDF-based data is machine-interpretable by design and inherently unambiguous. This is exploited in video indexing and retrieval by digital libraries, multimedia repositories (such as when searching for a particular procedure for medical training), knowledge discovery via inferring new statements automatically, and search engines, such as Google—think of the Knowledge Carousel and Knowledge Panels on Google's search engine result pages (SERPs), which retrieve information from, among others, structured data from website markup and LOD datasets.

Once identified, concepts can be interlinked with related data across LOD datasets, which can then be used for combined faceted and explorative video search [62]. In contrast to website contents retrieved through keyphrase-based web search, RDF-based knowledge representations can be queried and manipulated manually or pro-grammatically through the very powerful SPARQL query language [63]. SPARQL queries may include multiple questions in a single query to answer complex questions that cannot be formulated as keywords (which are used in traditional keyphrase-based web search). Furthermore, they can be executed not only on a single dataset, but also across multiple datasets using federated queries. For example, assume a task to retrieve three westerns starring Clint Eastwood that are shorter than 2.5 h, and order them alphabetically by title (see Listing 6.5).

Listing 6.8 Advanced querying with SPARQL

```
PREFIX dc: <http://purl.org/dc/terms/> .
PREFIX rdf: <http://www.w3.org/1999/02/22-rdf-syntax-ns#> .
PREFIX video: <http://purl.org/ontology/video/> .
PREFIX schema: <http://schema.org/> .
PREFIX xsd: <http://www.w3.org/2001/XMLSchema#> .
SELECT DISTINCT ?movie_title ?starring ?genre
FROM <http://example.com/sparql>
WHERE {
  ?video a schema:Movie ; dc:title ?movie_title ;
  video:starring ?starring ; schema:genre ?genre ;
   video:runningTime ?runningTime .
  FILTER (?starring = "Clint Eastwood"^^xsd:string) .
  FILTER (?genre = "western"^^xsd:string) .
  FILTER (?runningTime < "150"^^xsd:decimal) .
}
ORDER BY ?movie_title LIMIT 3
```

6.4.2   Video Semantics in Hypervideo Applications

Reasoning over structured video data can be used for, among other things, auto-matically generating annotations for constrained videos [64], providing context-awareness for augmented reality videos [65], semantically enriching interactive video playback [66], achieving collaborative annotation [67], and considering user preferences in video recommendation engines for video sharing portals, video libraries, and e-commerce.

6.5 Summary

Researched by both the signal processing and the knowledge engineering communities, the Semantic Gap in videos poses a real challenge. The formal representation of concepts depicted in videos and the rule-based description of video events are the knowledge engineering approaches that have already been successfully implemented for high-level concept mapping in constrained videos, such as medical videos, news videos, and sport videos. The description logic-based formal grounding of ontologies used for video representation ensures well-understood computational properties and decidability. Since the general-purpose description logics used in other fields lack the expressivity needed for representing complex video events, spatiotemporal annotations often employ spatial, temporal, and fuzzy description logics and rule-based mechanisms as well. The captured video semantics can be improved further via information fusion by taking into account low-level audio descriptors, and if available, metadata and subtitles. The high-level video semantics expressed using the above formalisms can be utilized in a range of intelligent applications from content-based video retrieval to hypervideo players. Knowledge discovery can be achieved via reasoning, for example previously unknown precursors of diseases might be automatically discovered via the co-occurrence of particular symptoms in medical videos.

The main challenges of using high-level descriptors in video scene understanding include the limitations of concept coverage of knowledge representations, the inherently ambiguous interpretations, and the reliable automation of structured video annotation.

References

1. Lowe, D.G.: Object recognition from local scale-invariant features. In: 7th IEEE International Conference on Computer Vision, Kerkyra, September 1999, vol. 2, pp. 1150–1157. IEEE, New York (1999). https://doi.org/10.1109/ICCV.1999.790410
2. Dollár, P., Rabaud, V., Cottrell, G., Belongie, S.: Behavior recognition via sparse spatio-temporal features. In: 2005 IEEE International Workshop on Visual Surveillance and Performance Evaluation of Tracking and Surveillance, Beijing, October 2005, pp. 65–72. IEEE, New York (2005). https://doi.org/10.1109/VSPETS.2005.1570899
3. Dalal, N., Triggs, B.: Histograms of oriented gradients for human detection. In: 2005 IEEE Computer Society Conference on Computer Vision and Pattern Recognition, San Diego, June 2005, vol. 1, pp. 886–893. IEEE Computer Society, Washington (2005). https://doi.org/10.1109/CVPR.2005.177
4. Dalal, N., Triggs, B., Schmid, C.: Human detection using oriented histograms of flow and appearance. In: Leonardis, A., Bischof, H., Pinz, A. (eds.) Computer Vision—ECCV 2006.

9th European Conference on Computer Vision, Graz, May 2006. Lecture Notes in Computer Science, vol. 3952, pp. 428–441. Springer, Heidelberg (2006). https://doi.org/10.1007/11744047_33

5. Bay, H., Ess, A., Tuytelaars, T., Van Gool, L.: Speeded-up robust features (SURF). Comput. Vis. Image. Und. **110**(3), 346–359 (2008). https://doi.org/10.1016/j.cviu.2007.09.014

6. Xu, F., Zhang, Y-J.: Evaluation and comparison of texture descriptors proposed in MPEG-7. J. Vis. Commun. Image Rep. **17**(4), 701–716 (2006). https://doi.org/10.1016/j.jvcir.2005.10.002

7. Yang, N.-C., Chang, W.-H., Kuo, C.-M., Li, T.-H.: A fast MPEG-7 dominant color extraction with new similarity measure for image retrieval. J. Vis. Commun. Image Rep. **19**(2), 92–105 (2008). https://doi.org/10.1016/j.jvcir.2007.05.003

8. Viola, P., Jones, M.: Rapid object detection using a boosted cascade of simple features. In: IEEE Computer Society Conference on Computer Vision and Pattern Recognition, Kauai, Dec 8–14, 2001, pp. 511–518 (2001). https://doi.org/10.1109/CVPR.2001.990517

9. Lienhart, R., Maydt, J.: An extended set of Haar-like features for rapid object detection. In: International Conference on Image Processing, Rochester, September 2002, pp. 900–903 (2002). https://doi.org/10.1109/ICIP.2002.1038171

10. Lowe, D.G.: Distinctive image features from scale-invariant keypoints. Int. J. Comput. Vision **60**(2), 91–110 (2004). https://doi.org/10.1023/B:VISI.0000029664.99615.94

11. Khedher, M.I., El Yacoubi, M.A.: Local sparse representation based interest point matching for person re-identification. In: Arik, S., Huang, T., Lai, W.K., Liu, Q. (eds.) Neural Information Processing. 22nd International Conference on Neural Information Processing, Turkey, November 2015. Lecture Notes in Computer Science, vol. 9491, pp. 241–250. Springer, Cham (2015). https://doi.org/10.1007/978-3-319-26555-1_28

12. Rublee, E., Rabaud, V., Konolige, K., Bradski, G.: ORB: an efficient alternative to SIFT or SURF. In: 2011 IEEE International Conference on Computer Vision, Barcelona, Nov 6–13, 2011, pp. 2564–2571 (2011). https://doi.org/10.1109/ICCV.2011.6126544

13. Sikos, L.F.: Description logics in multimedia reasoning. Springer, Cham (2017). https://doi.org/10.1007/978-3-319-54066-5

14. Boll, S., Klas, W., Sheth, A.: Overview on using metadata to manage multimedia data. In: Sheth, A., Klas, W. (eds.) Multimedia Data Management: Using Metadata to Integrate and Apply Digital Media, p. 3. McGraw-Hill, New York (1998)

15. Duong, T.H., Nguyen, N.T., Truong, H.B., Nguyen, V.H.: A collaborative algorithm for semantic video annotation using a consensus-based social network analysis. Expert. Syst. Appl. **42**(1), 246–258 (2015). https://doi.org/10.1016/j.eswa.2014.07.046

16. Ballan, L., Bertini, M., Del Bimbo, A., Seidenari, L., Serra, G.: Event detection and recognition for semantic annotation of video. Multimed. Tools Appl. **51**(1), 279–302 (2011). https://doi.org/10.1007/s11042-010-0643-7

17. Gómez-Romero, J., Patricio, M.A., García, J., Molina, J.M.: Ontology-based context representation and reasoning for object tracking and scene interpretation in video. Expert. Syst. Appl. **38**, 7494–7510 (2010). https://doi.org/10.1016/j.eswa.2010.12.118

18. Poppe, C., Martens, G., De Potter, P., Van de Walle, R.: Semantic web technologies for video surveillance metadata. Multimed. Tools Appl. **56**(3), 439–467 (2012). https://doi.org/10.1007/s11042-010-0600-5

19. Bohlken, W., Neumann, B., Hotz, L., Koopmann, P.: Ontology-based realtime activity monitoring using beam search. In: Crowley, J.L., Draper, B.A., Thonnat, M. (eds.) Computer Vision Systems. ICVS 2011. Lecture Notes in Computer Science, vol. 6962, pp. 112–121. Springer, Heidelberg (2011). https://doi.org/10.1007/978-3-642-23968-7_12

20. Wu, Z., Yao, T., Fu, Y., Jiang, Y.-G.: Deep learning for video classification and captioning (2016). arXiv:1609.06782

21. Herrera, J.L., del-Blanco, C.R., García, N.: Improved 2D-to-3D video conversion by fusing optical flow analysis and scene depth learning. In: 3DTV-Conference: The True Vision—Capture, Transmission and Display of 3D Video, Hamburg, June 2016. IEEE, New York (2016). https://doi.org/10.1109/3DTV.2016.7548954

22. Sikos, L.F.: A novel ontology for 3D semantics: ontology-based 3D model indexing and content-based video retrieval applied to the medical domain. Int. J. Metadata Semant. Ontol. 12(1), 59–70 (2017). https://doi.org/10.1504/IJMSO.2017.10008658

23. Gruber, T.R.: Towards principles for the design of ontologies used for knowledge sharing. In: Guarino, N., Poli, R. (eds.) Formal Ontology in Conceptual Analysis and Knowledge Representation. Kluwer Academic Publishers, Deventer (1993)

24. Perperis, T., Giannakopoulos, T., Makris, A., Kosmopoulos, D.I., Tsekeridou, S., Perantonis, S.J., Theodoridis, S.: Multimodal and ontology-based fusion approaches of audio and visual processing for violence detection in movies. Expert Syst. Appl. 38(11), 14102–14116 (2011). https://doi.org/10.1016/j.eswa.2011.04.219

25. Rodríguez-García, M.Á., Colombo-Mendoza, L.O., Valencia-García, R., Lopez-Lorca, A.A., Beydoun, G.: Ontology-based music recommender system. In: Omatu, S., Malluhi, Q.M., Gonzalez, S.R., Bocewicz, G., Bucciarelli, E., Giulioni, G., Iqba, F. (eds.) 12th International Conference on Distributed Computing and Artificial Intelligence, Salamanca, June 2015. Advances in Intelligent Systems and Computing, vol. 373, pp. 39–46. Springer, Cham (2015). https://doi.org/10.1007/978-3-319-19638-1_5

26. Sikos, L.F.: A novel approach to multimedia ontology engineering for automated reasoning over audiovisual LOD datasets. In: Nguyễn, N.T., Trawiński, B., Fujita, H., Hong, T.-P. (eds.) Intelligent Information and Database Systems. 8th Asian Conference on Intelligent Information and Database Systems, Đà Nẵng, March 2016. Lecture Notes in Computer Science (Lecture Notes in Artificial Intelligence), vol. 9621, pp. 3–12. Springer, Heidelberg (2016). https://doi.org/10.1007/978-3-662-49381-6_1

27. Davis, S., Burnett, I., Ritz, C.: Using social networking and collections to enable video semantics acquisition. IEEE MultiMedia PP(99). https://doi.org/10.1109/MMUL.2009.72

28. Bertini, M., Del Bimbo, A., Torniai, C.: Automatic annotation and semantic retrieval of video sequences using multimedia ontologies. In: MM 2006 Proceedings of the 14th ACM International Conference on Multimedia, Santa Barbara, October 2006, pp. 679–682. ACM, New York (2006)

29. Gómez-Romero, J., García, J., Patricio, M.A., Serrano, M.A., Molina, J.M.: Context-based situation recognition in computer vision systems. In: Gómez-Romero, J., García, J., Patricio, M.A., Serrano, M.A., Molina, J.M. (eds.) Context-enhanced Information Fusion. Advances in Computer Vision and Pattern Recognition, pp. 627–651. Springer, Cham (2016). https://doi.org/10.1007/978-3-319-28971-7_23

30. Sikos, L.F.: Mastering Structured Data on the Semantic Web: From HTML5 Microdata to Linked Open Data. Apress, New York (2015). https://doi.org/10.1007/978-1-4842-1049-9_1

31. Krötzsch, M., Simančík, F., Horrocks, I.: A description logic primer (2013). arXiv:1201.4089v3

32. Sikos, L.F.: Web Standards: Mastering HTML5, CSS3, and XML, 2nd ed. Apress, New York (2014). https://doi.org/10.1007/978-1-4842-0883-0

33. Isaac, A., Troncy, R.: Designing and using an audio-visual description core ontology. Paper presented at the Workshop on Core Ontologies in Ontology Engineering, Northamptonshire, 8 (2004). (Oct)

34. Hunter, J.: Adding multimedia to the Semantic Web—building an MPEG-7 ontology. Presented at the 1st International Semantic Web Working Symposium, Stanford University, Stanford, 29 July–1 Aug 2001

35. Tsinaraki, C., Polydoros, P., Moumoutzis, N., Christodoulakis, S.: Integration of OWL ontologies in MPEG-7 and TV-Anytime compliant semantic indexing. In: Persson, A., Stirna, J. (eds.) Advanced Information Systems Engineering. 16th International Conference on Advanced Information Systems Engineering, Riga, June 2004. Lecture Notes in Computer Science, vol. 3084, pp. 398–413. Springer, Heidelberg (2004). https://doi.org/10.1007/978-3-540-25975-6_29

36. García, R., Celma, O.: Semantic integration and retrieval of multimedia metadata. Paper presented at the 5th International Workshop on Knowledge Markup and Semantic Annotation, Galway, 7 Nov 2005
37. Blöhdorn, S., Petridis, K., Saathoff, C., Simou, N., Tzouvaras, V., Avrithis, Y., Handschuh, S., Kompatsiaris, Y., Staab, S., Strintzis, M.: Semantic annotation of images and videos for multimedia analysis. In: Gómez-Pérez, A., Euzenat, J. (eds.) The Semantic Web: Research and Applications. Second European Semantic Web Conference, Heraklion, May–June 2005. Lecture Notes in Computer Science, vol. 3532, pp. 592–607. Springer, Heidelberg (2005). https://doi.org/10.1007/11431053_40
38. Athanasiadis, T., Tzouvaras, V., Petridis, K., Precioso, F., Avrithis, Y., Kompatsiaris, Y.: Using a multimedia ontology infrastructure for semantic annotation of multimedia content. In: Paper presented at the 5th International Workshop on Knowledge Markup and Semantic Annotation, Galway, 7 Nov 2005
39. Oberle, D., Ankolekar, A., Hitzler, P., Cimiano, P., Sintek, M., Kiesel, M., Mougouie, B., Baumann, S., Vembu, S., Romanelli, M.: DOLCE ergo SUMO: on foundational and domain models in the SmartWeb integrated ontology (SWIntO). J. Web Semant. Sci. Serv. Agents World Wide Web 5(3), 156–174 (2007). https://doi.org/10.1016/j.websem.2007.06.002
40. Dasiopoulou, S., Tzouvaras, V., Kompatsiaris, I., Strintzis, M.: Capturing MPEG-7 semantics. In: Sicilia, M.-A., Lytras, M.D. (eds.) Metadata and Semantics, pp. 113–122. Springer, New York (2009)
41. Arndt, R., Troncy, R., Staab, S., Hardman, L.: COMM: a core ontology for multimedia annotation. In: Staab, S., Studer, R. (eds.) Handbook on Ontologies, pp. 403–421, Springer, Heidelberg (2009). https://doi.org/10.1007/978-3-540-92673-3_18
42. Sikos, L.F.: 3D model indexing in videos for content-based retrieval via X3D-based semantic enrichment and automated reasoning. In: 22nd International Conference on 3D Web Technology, Brisbane, June 2017. ACM, New York (2017). https://doi.org/10.1145/3055624.3075943
43. Naphade, M., Smith, J.R., Tesic, J., Chang, S.-F., Hsu, W., Kennedy, L., Hauptmann, A., Curtis, J.: Large-scale concept ontology for multimedia. IEEE MultiMedia 13(3), 86–91 (2006). https://doi.org/10.1109/MMUL.2006.63
44. Zha, Z.-J., Mei, T., Zheng, Y.-T., Wang, Z., Hua, X.-S.: A comprehensive representation scheme for video semantic ontology and its applications in semantic concept detection. Neurocomputing 95, 29–39 (2012). https://doi.org/10.1016/j.neucom.2011.05.044
45. Hogenboom, F., Borgman, B., Frasincar, F., Kaymak, U.: Spatial knowledge representation on the Semantic Web. In: 2010 IEEE Fourth International Conference on Semantic Computing (2010). https://doi.org/10.1109/ICSC.2010.31
46. Yildirim, Y., Yazici, A., Yilmaz, T.: Automatic semantic content extraction in videos using a fuzzy ontology and rule-based model. IEEE Trans. Knowl. Data Eng. 25(1), 47–61 (2013). https://doi.org/10.1109/TKDE.2011.189
47. Sikos, L.F.: Spatiotemporal Reasoning for Complex Video Event Recognition in Content-Based Video Retrieval. In: Hassanien, A.E., Shaalan, K., Gaber, T., Tolba, M. (eds.) 3rd International Conference on Advanced Intelligent Systems and Informatics, Cairo, September 2017. Advances in Intelligent Systems and Computing, vol. 639, pp. 704–713. Springer, Cham (2017). https://doi.org/10.1007/978-3-319-64861-3_66
48. Sikos, L.F., Powers, D.M.W.: Knowledge-driven video information retrieval with LOD: from semi-structured to structured video metadata. In: 8th Workshop on Exploiting Semantic Annotations in Information Retrieval, Melbourne, October 2015. pp. 35–37. ACM, New York (2015). https://doi.org/10.1145/2810133.2810141
49. Saatho, C., Scherp, A.: M3O: The multimedia metadata ontology. Presented at the 10th International Workshop of the Multimedia Metadata Community on Semantic Multimedia Database Technologies, Graz, 2 Dec 2009
50. Horvat, M., Bogunović, N., Ćosić, K.: STIMONT: a core ontology for multimedia stimuli description. Multimed. Tools Appl. 73(3), 1103–1127 (2014). https://doi.org/10.1007/s11042-013-1624-4

51. Sikos, L.F.: Advanced (X)HTML5 metadata and semantics for Web 3.0 videos. DESIDOC J. Library Inf. Technol. **31**(4), 247–252 (2011). https://doi.org/10.14429/djlit.31.4.1105
52. Bizer, C., Heath, T., Berners-Lee, T.: Linked Data—the story so far. Semant. Web Inform. Syst. **5**(3), 1–22 (2009). https://doi.org/10.4018/jswis.2009081901
53. Choudhury, S., Breslin, J.G., Passant, A.: Enrichment and ranking of the YouTube tag space and integration with the Linked Data Cloud. In: The Semantic Web—ISWC 2009. 8th International Semantic Web Conference, Chantilly, October 2009. Lecture notes in computer science, vol. 5823, pp. 747–762. Springer, Heidelberg (2009). https://doi.org/10.1007/978-3-642-04930-9_47
54. Sikos, L.F.: RDF-powered semantic video annotation tools with concept mapping to Linked Data for next-generation video indexing. Multimed. Tools Appl. **76**(12), 14437–14460 (2016). https://doi.org/10.1007/s11042-016-3705-7
55. Jiang, Y.-G., Bhattacharya, S., Chang, S.-F., Shah, M.: High-level event recognition in unconstrained videos. Int. J. Multimed. Inf. Retrieval **2**(2), 73–101 (2013). https://doi.org/10.1007/s13735-012-0024-2
56. Elleuch, N., Zarka, M., Ammar, A.B., Alimi, A.M.: A fuzzy ontology-based framework for reasoning in visual video content analysis and indexing. In: Eleventh International Workshop on Multimedia Data Mining, San Diego, Aug 21–24, 2011, Article 1 (2011). https://doi.org/10.1145/2237827.2237828
57. Allen, J.F.: Maintaining knowledge about temporal intervals. Commun. ACM **26**(11), 832–843 (1983). https://doi.org/10.1145/182.358434
58. Bai, L., Lao, S., Zhang, W., Jones, G.J.F., Smeaton, A.F.: Video semantic content analysis framework based on ontology combined MPEG-7. In: Boujemaa, N., Detyniecki, M., Nürnberger, A. (eds.) Adaptive Multimedia Retrieval: Retrieval, User, and Semantics. 5th International Workshop on Adaptive Multimedia Retrieval, Paris, July 2007. Lecture Notes in Computer Science, vol. 4918, pp. 237–250. Springer, Heidelberg (2008). https://doi.org/10.1007/978-3-540-79860-6_19
59. Bertini, M., Del Bimbo, A., Serra, G.: Video event annotation using ontologies with temporal reasoning. In: Proceeding of the 2nd DELOS Conference, Padova, January 2008, pp. 13–23 (2008)
60. Stoilos, G., Stamou, G., Pan, J.Z.: Fuzzy extensions of OWL: logical properties and reduction to fuzzy description logics. Int. J. Approximate Reasoning **51**(6), 656–679 (2010). https://doi.org/10.1016/j.ijar.2010.01.005
61. Zarka, M., Ammar, A.B., Alimi, A.M.: Fuzzy reasoning framework to improve semantic video interpretation. Multimed. Tools Appl. **75**(10), 5719–5750 (2015). https://doi.org/10.1007/s11042-015-2537-1
62. Waitelonis, J., Sack, H.: Towards exploratory video search using Linked Data. Multimed. Tools Appl. **59**(2), 645–672 (2012). https://doi.org/10.1007/s11042-011-0733-1
63. Lee, M.-H., Rho, S., Choi, E.-I.: Ontology-based user query interpretation for semantic multimedia contents retrieval. Multimed. Tools Appl. **73**(2), 901–915 (2014). https://doi.org/10.1007/s11042-013-1383-2
64. Ballan, L., Bertini, M., Del Bimbo, A., Serra, G.: Semantic annotation of soccer videos by visual instance clustering and spatial/temporal reasoning in ontologies. Multimed. Tools Appl. **48**(2), 313–337 (2010). https://doi.org/10.1007/s11042-009-0342-4
65. Münzer, B., Schoeffmann, K., Böszörményi, L.: Content-based processing and analysis of endoscopic images and videos: a survey. Multimed Tools Appl. (2017). https://doi.org/10.1007/s11042-016-4219-z
66. Nixon, L., Bauer, M., Bara, C., Kurz, T., Pereira, J.: ConnectME: semantic tools for enriching online video with web content. In: Proceedings of the 8th International Conference on Semantic Systems, Graz, Austria (2012)
67. Grassi, M., Morbidoni, C., Nucci, M.: A collaborative video annotation system based on semantic web technologies. Cogn. Comput. **4**(4), 497–514 (2012). https://doi.org/10.1007/s12559-012-9172-1

Chapter 7
Deep Learning—A New Era in Bridging the Semantic Gap

Urszula Markowska-Kaczmar and Halina Kwaśnicka

Abstract The chapter deals with the semantic gap, the well-known phenomenon in the area of vision systems. Despite the significant efforts of researchers, the problem of how to overcome the semantic gap remains a challenge. One of the most popular research areas, where this problem is present and causes difficulty in obtaining good results, is the task of image retrieval. This chapter focuses on this problem. As deep learning models gain more and more popularity among researchers and more and more spectacular results are obtained, the deep learning models in solving the semantic gap in the Content Based Image Retrieval (CBIR) is the central issue of this chapter. The chapter briefly presents the traditional approaches to CBIR, next introduces shortly into methods and models of deep learning, and shows the application of deep learning at the particular levels of CBIR—features level, common sense knowledge level, and inference about a scene level.

7.1 Introduction

Image processing is an important and current research area from the scientific and practical point of view. Different image processing techniques are used in such areas as medicine, astronomy, archeology, electronic games, video communications, and others. Extraction of useful information from the processed image is an essential task in image processing, some of such methods try to mimic human visual processes. By object recognition, we can obtain information about the names of all or some of objects in the image. However, how to possess the semantic knowledge from images is still the unsolved scientific problem. Apparently similar images can hide a

U. Markowska-Kaczmar (✉)
Wroclaw University of Science and Technology, Wyb. Wyspianskiego 27,
50-370 Wroclaw, Poland
e-mail: urszula.markowska-kaczmar@pwr.edu.pl

H. Kwaśnicka
Depatment of Computational Intelligence, Wroclaw University
of Science and Technology, Wroclaw, Poland
e-mail: halina.kwasnicka@pwr.edu.pl

© Springer International Publishing AG, part of Springer Nature 2018
H. Kwaśnicka and L. C. Jain (eds.), *Bridging the Semantic Gap in Image
and Video Analysis*, Intelligent Systems Reference Library 145,
https://doi.org/10.1007/978-3-319-73891-8_7

123

semantically different content and vice versa what causes strong difficulties in image analysis. People can interpret images according to the context by inference and use some prior knowledge. This problem is seen in the image retrieval area, where the similarity of images plays an important if not a crucial role. The term similarity is imprecise, in fact, it is very subjective. The same image can be interpreted differently by different people depending on their socio-cultural background, usage purpose, and contextual background.

Content Based Image Retrieval (CBIR) is a technique of searching images according to a user's interest on the basis of visual features extracted from the image, usually large scale image databases are searched. The image retrieving is a challenge when we expect the results according to human perspective and expectations. The difference between the low-level representation of the image and its high-level human perception is known as a semantic gap. Even the semantic gap has been extensively discussed in the literature [42, 43, 189, 201] it remains an open problem. Datta et al. [33] minded the role of the higher-level perception, (they denote it as aesthetics). It concerns a kind of emotions a picture arouses in people and adds a new dimension to image understanding, benefiting CBIR. Smeulders et al. [189] define the semantic gap within CBIR as "the lack of coincidence between the information that one can extract from the visual data acquired from an image and the interpretation that the same data have for a user in a given situation." It manifests as the difference between user intent and the content of returned images. The user intent can be defined as a query by example, in this case, the problem lies in image descriptors matching, or a query can be expressed in formal/natural language—it refers to the captions or labels analysis. Usually, a user seeks for semantic similarity, while in many cases the CBIR system considers similarity only by visual analysis. The content of an image is identified on the low-level pixel data. A linguistic description of an image, even it is not always precise, is more contextual than raw visual data. Therefore the most immediate means to embed semantic characteristics of an image is to entail an image with annotation by keywords or captions. It is an old concept that can reduce content-based access to information retrieval [174]. Nonetheless, research suggests it could be beneficial to use features from both sources: visual and linguistic (a multi-modal search). In practice, labeling images is expensive and context-sensitive.

Deep Learning (DL) is a new area of the Machine Learning domain. It was developed from 80-ties of 20 century [87, 117] but its popularity started in 2006 when Hinton et al. [83] showed how multilayered feedforward neural network could be pre-trained efficiently, layer by layer, treating each layer in turn as an unsupervised restricted Boltzmann machine. Then fine-tuning the whole multilayered structure uses supervised backpropagation [84]. Now the list of DL applications is impressive, starting with computer vision and pattern recognition [23, 51, 63, 99, 105, 152, 168], through computer games, robots and self-driving cars [18, 115, 144, 198], voice recognition and generation [7, 86], music composition [32, 96], transferring style from famous paintings [25, 55], and ending with automatic translation [56, 221]—only to give some examples.

Taking into account that CBIR is open research problem in terms of semantic gap bridging, we focus on this task. From the other hand, Deep Learning (DL) is a new approach in Machine Learning. In recent years DL has been applied to hundreds of problems, including computer vision and natural language processing, in both academia and the industry. The main goal of this chapter is to present short survey of using DL in bridging the semantic gap in CBIR area.

The rest of the chapter is organized as follows. In the next section, we shortly present the development of CBIR approaches. Section three introduces the Deep Learning paradigm and main deep architectures used in the CBIR problem. In this section, we also briefly mention how DL is used for two important from the CBIR point of view tasks, namely—visual attention modeling, and embedding semantic features (word encoding and language model building). The last is essential when someone wants to join visual and text information into CBIR systems. Section four presents the usefulness of DL on three levels of basic structure for scene interpretation with a DL system, i.e.: (i) low-level, called feature level, its aim is a description of the image with a set of (the best) features; (ii) the second level, in which the common sense knowledge is built by learning temporal and spatial knowledge from aligned visual and textual data; (iii) the third level, it contains an inference about the scene based on the second level. The last section concludes the chapter.

7.2   Content Based Image Retrieval

First systems for image retrieval used text describing an image content, and they exploited text retrieval techniques. These search engines used the text manually associated with an image or automatically extracted from tags or captions of Web images. Annotation process is time-consuming and labor-intensive. Another disadvantage of this approach stems from ambiguous or even irrelevant words assigned to the image and difficulties in using another language description because the query has to meet the language of surrounding text. The surrounding text only partially describes the semantic content of images, and sometimes the results are poor [93]. The reason lies in polysemy—a phenomenon when one word can have different meaning [176]. Another reason is human's perception subjectivity. In response to the described problems, Content Based Image Retrieval (CBIR) systems have arrived, where the query is given by using an image or sketch. Initially, the CBIR term [40] has been used to describe the process of retrieving desired images from an image collection based on features automatically extracted from the images. In early CBIR systems the visual features played essential role [210], they included colors, shapes, and textures. The first global features used in CBIR systems have required relatively low computational cost, however they were not invariant to image transformations. The further research were focused on the design of invariant and discriminative features—local features. The advantage of local features are their robust to occlusions, cropping and geometric transformations [202]. The popular detectors are corner detectors: Harris, Shi-Tomasi and FAST [73, 175, 186], and blob detectors: SIFT [6, 127] and SURF [14]. A comparative survey of local descriptors has been presented in [142]. Features are used to build Bag of Words (BoW) model borrowed from the text analysis [3]. The similarity between BoW vectors can be computed by the standard similarity distances, such as the Euclidian or Manhattan measures.

Eakins [41] distinguishes between various levels of image retrieval. For this survey, the semantic level that requires the identification of images on the basis of desired types of object, scene, event, or abstract ideas, is the most important. Image semantics recognition is a challenge, because it is not only hidden inside the image but depends on a priori knowledge and the user objectives. Progress in CBIR is closely interrelated with the development of new methods for image matching, image recognition, image segmentation, image annotation and object detection [40]. In the context of the semantic gap, the crucial problems are: understanding of users' intention, developing the appropriate description of image content, automatic extraction of features that in the best way represent image content, and images matching that reflects human similarity judgments. In image matching different similarity can be used, their comparative studies are in [147]. Summing up, the early years of CBIR development we can say that research was focused mainly on object level retrieval, not on the semantic level of an image.

The semantic gap problem was noticed, and it received a considerable attention of researchers in the last decade. Developed systems use not only visual information extracted from an image but also other features, e.g., text. A fusion of the two basic image modalities—text (usually represented by keywords or captions) and visual features is very promising to bridge the semantic gap [75]. Feature representation with embedded semantics, i.e., including textual features, is learned using probabilistic Latent Semantic Analysis (pLSA) model [89] and Latent Dirichlet Allocation (LDA) model [17]. Both models have been adopted to CBIR problem [119]. Piras et al. [165] underline that the subjectivity of image semantics adaptation needs a fusion of different image representations. They give a comprehensive survey of existing fusion methods. They distinguish two groups: early and late fusion. Early fusion refers to the combination of the features into a single representation before the computation of similarity between images. This approach includes concatenation of various descriptors or different feature spaces. Late fusion is compound either of the outputs produced by different retrieval systems or of the similarity rankings, the outputs and the rankings referring to different feature representations [45].

It is worth mentioning the idea of saliency and attention. Saliency [163] tries to mimic how a human eye identifies important objects on the scene and is based on a simple fundamental element—a contrast between an object and its neighbor. The saliency model guides vision to potentially meaningful parts of a scene. Most of the saliency detection methods use only the low-level image features, e.g., contrast, edge, intensity. So, it is difficult to capture the task-specific semantic information. Predicting saliency points can be applied to object detection [20, 26], unsupervised object discovery and classification [236]. Attention is defined as the selection the most relevant regions of a scene. These regions contain essential visual concepts. Based on sparse representation, Bruce et al. [21] proposed the Attention by Information Maximization (AIM) model which adopts the self-information of Independent Component Analysis (ICA) coefficients as the measure for signal saliency. In the paper [22], Bruce et al. discussed some challenges faced in models of visual saliency. Based on information theory and sparse coding, Hou et al. [92] proposed the Dynamic Visual Attention (DVA) model which defines spatiotemporal saliency as incremental coding length. Garcia-Diaz et al. [53] proposed the Adaptive Whitening Saliency (AWS) model which relies on a contextually adapted representation produced through adaptive whitening of color and scale features. An extensive comparative study of visual saliency models, belonging to the majority class of models, is presented in [19].

Another trend in CBIR in the last years explores test-bed ontologies combined with content based techniques and annotation to narrow the semantic gap [71]. Hare et al. distinguish two possible approaches in attacking the semantic gap: from bellow—analysing the gap from descriptors to labels (by auto-annotation and/or semantic spaces) and from above—analysing the gap from labels to semantics (looking at the use of ontologies). To bridge the gap from below, Hare et al. [72] propose a new technique for auto-annotation.

One can find many papers reviewing the semantic gap problem. A comprehensive survey presented in [125] identifies various directions to narrow down the semantic gap (using an ontology, machine learning, generating the semantic template, fusing text and visual content of images). Image semantics is widely concerned in another review paper [199]. The authors distinguish various methods used for the semantic analysis of images. They deeply discuss: *direct* methods using a plain representation of data and plain statistical methods; *linguistic* methods using an intermediate visual vocabulary between raw numerical data and high-level semantics; *compositional* methods where parts of the image are identified in the segmentation process before the whole image or its parts are annotated; *structural* methods where a geometry of image parts is used; *hierarchical compositional* methods that construct a hierarchy of parts; *communicating* methods when information is shared between categories; *hierarchical* methods that search for hierarchical relationships between categories; *multi-label* methods assigning several global labels simultaneously to an image.

Alzu'bi et al. in [3] noticed the role of deep learning in CBIR and bridging the semantic gap. They focus on the role of the convolutional network in visual feature extraction and shortly describe image captioning method from [105]. Zhou et al. [235] extend an overview of this subject in many aspects, for instance, by describing a generation of patch-level feature representation based on convolutional kernel networks [159] and deep network for hashing images into short binary codes with optimization based on triplet ranking loss [114].

7.3   Deep Learning

In recent years Deep Learning (DL) is growing in the number of new concepts and the number of successful challenging applications. We do not cite here the precise definition of Deep Learning. Goodfellow et al. explain this term in a descriptive way [65]. They underline that solving the tasks that are easy for people to perform but hard for people to describe them formally is a challenge to Artificial Intelligence.

As the examples of such problem, we can mention understanding spoken words or recognition of faces on images. People solve such kind of problems more intuitively. Approaches in which solutions of such problems are searched by learning from experience, by defining a concept through its relation to simpler concepts, is called deep learning. A graph that shows how the concepts are built on top of each other is deep, it contains many layers, therefore this approach is called as *deep learning*. Deep learning avoids the need for people to specify all the knowledge that the computer can need to solve such a problem. In the mid-1980s, Hinton and others propagated neural networks with 'deep' models, consisting of many layers, but training them required heavy human involvement, e.g., labeling enormous data sets, and there was not available required computational power for such complicated tasks as speech or image recognition. In 2006, Hinton and researchers developed another way of training deep models, by teaching four individual layers of neurons. In such a deep network, higher levels capture more abstract concepts. The big advantage of DLs is that they can discover features that in the best way represent the problem. Shallow neural networks start with handcrafted features of the image, but deep learning starts with the raw pixels and learns features automatically during the training process, from primitive features to more abstract in successive layers. The DL can be defined as a part of Machine Learning consisting of algorithms used to model high-level abstractions in data using architectures composed of multiple nonlinear transformations.

The deep models, consisting of a number of layers, need a lot of data and computation power and time to train. They owe its quick development to the growth of unannotated data amount and acceleration of computation by using GPUs. Also, new techniques that allow to train networks more effectively have been proposed: the mini batch training (the batch algorithm keeps the system weights constant while computing the error associated with each sample in the input) [15, 85, 98], new optimization algorithms based on back-propagated gradient and gradient-based optimization [16, 39, 108, 195, 230, 231], dropout [192], regularization [65].

Shallow networks typically used sigmoid transfer functions or hyperbolic tangent. While applying to deep networks, these transfer functions cause vanishing gradient during training. New transfer functions, like ReLU or Leaky ReLU and others [76, 131] reduce vanishing gradient, cause sparsity and make training faster.

While deep learning is computationally demanded and needs lot of training time, it is valuable that some of the trained deep models are archived with all parameters. They are ready to adapt and train for other related tasks which have too few training examples to learn a full deep representation. In the fine-tuning step, some weights are frozen, they are good starting points to further adjust weights. This phenomenon is called transfer learning. It is very popular in an image recognition problem [2, 55, 91, 225]. It is really valuable in medical image recognition [138, 173], where the data sets are usually small.

One way to improve efficiency training is to use a big dataset. When the dataset is small, the solution lies in data augmentation [6]. This technique is widely applied in deep learning research.

7.3.1 Deep Learning Architecture

The following subsections briefly introduce into deep models that could be crucial in reference to bridging the semantic gap in the image retrieval.

7.3.1.1 Autoencoder and Stacked Autoencoders

Autoencoder is a general idea of the network with one hidden layer that is trained in the reconstruction mode. This means that the network is trained in unsupervised way using standard gradient optimization method trying to reconstruct its input on the output neurons. During the training process, the activations of hidden neurons search for a latent representation of the problem, i.e., they automatically find features (code-word). This effect is very beneficial in relation to the handcrafted feature development and can be used to reduce the dimensionality of input data [82]. The simplest form is based on the MultiLayer Perceptron (MLP) network [74]. Each neuron sums weighted input signals (weights are the parameters searched during training) and transform it by an activation function. The vanishing gradient problem arrives, when sigmoid activation function is applied. Therefore, with growing interest in deep learning, novel activation functions have been proposed: [1, 29, 59, 101, 130]. Autoencoder consists of two parts. The first one is an encoder and the second one a decoder.

Usually, the MLP uses a loss function which is used to measure the inconsistency between predicted value $W''H$ and actual input X. The most popular is an error between the input vector X and the output vector $W''H$.

$$\mathcal{L} = ||W''H - X||_2^2 \tag{7.1}$$

where W'' is a weights matrix in the decoding part, H is the activations vector of a hidden layer and X is an input vector. Minimising the squared error is equivalent to predicting the (conditional) mean of the output.

There are many different autoencoders. Sparse autoencoders are a type of autoencoder enforced to learn a code dictionary (feature representation) that minimizes reconstruction error while restricting the number of code-words required for reconstruction.

Sparsity may be achieved by additional terms $\lambda||H||_1$ in the loss function during training [150] (by comparing the probability distribution of the hidden unit activations with some low desired value).

Denoising autoencoders [205] take a partially corrupted input while training to recover the original undistorted input.

Stacking autoencoders (deep autoencoders) use greedy layer-wise training as pretraining. Each layer in the network learns an encoding of the layer below. Then the network is fine-tuned. In this way, a network can learn hierarchical features in an unsupervised manner (Vincent et al. [206]). Next, it is finetuned by training in the supervised way. To use the network as a classifier, decoding part is ignored and the layer of softmax neurons is added. The features from encoder can also be useful for clustering.

The very interesting structure is Variational Autoencoder (VAE) [107, 172]. It recovers the data distribution and learns latent features. After training it can be used as a generative model.

7.3.1.2   Probabilistic Graphical Models

The fundamental block in the probabilistic graphical model is Restricted Boltzmann Machine (RBM). The RBM consists of two-layers, one visible and one hidden layer. The visible layer takes an input, and after one cycle the output arrives at the visible layer. In some way, the architecture is similar to the feed-forward neural network with logistic activation function, although, training is different. Typically, training uses contrastive divergence (CD) [82, 85] with persistent Markov chains.

A Deep Belief Network (DBN) is formed by training RBMs one at a time and then stacking them on top of each other. Each hidden layer creates feature detectors. In such stack, features learned by one RBM are used as the input for training the next RBF. The idea is similar to deep autoencoder, but the interpretation is different. Once the RBM is stacked, it changes the prior distribution over the hidden values of the lower RBM in the stack. DBNs can be used to initialize a deep neural network that is easily fine-tuned by backpropagation. Applications of DBNs include natural language processing [181], speech recognition [146] and classification problems.

7.3.1.3   Convolutional Networks

Convolutional Network (CNN) is a feed-forward network that can extract topological features directly from pixels of an image. Similarly to almost all other neural networks, they are trained with a version of the back-propagation algorithm. The net-

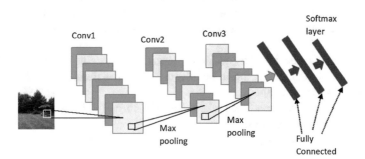

Fig. 7.1 Convolutional Neural Network (CNN)

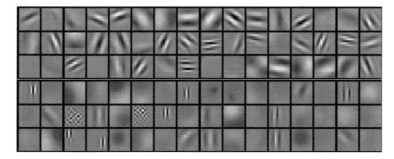

Fig. 7.2 Features of the first convolutional layer in CNN trained on ImageNet dataset [113]. Used with kind permission of Alex Krizhevsky

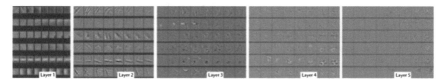

Fig. 7.3 Features in the subsequent layers of convolutional network [229]. Used with permission provided by Springer

work consists of various convolutional layers, pooling layers, and fully connected layers (Fig. 7.1). Convolutional layers are composed of feature maps. Each neuron in a feature map looks for the same feature but at different positions of the input image. All neurons in one feature map share the same weights. Each map is subsampled typically with mean or max pooling in pooling layers. Pooling provides a form of translation invariance. The intuition behind this layer is that once we know that a specific feature is in the original input volume, its exact location is not as important as its relative location to other features. The architecture of CNNs exploits spatially local connectivity between neurons of adjacent layers: each neuron is connected to only a small region of the input volume. In the case of classification, the last layer operates as the softmax layer.

CNNs trained on natural images learn on its first layer features similar to Gabor filters and color blobs (Fig. 7.2).

Features produced by first-layer are general, not specific to the particular dataset, it means that they can be applied to many datasets and tasks. Many pre-trained networks are archived. It is very popular to use them initializing a network by transferring features from almost any number of layers. Yosinski et al. showed that this approach could produce a boost to the network generalization [225]. Examples of pre-trained off-the-shelf convolutional networks are VGG-16, ResNet-50, AlexNet, Inception-ResNet-v1, GoogLeNet, LeNet. The networks were trained on more than a million images. Such trained model has learned rich feature representations for a wide range of images [187]. In the subsequent layers learned features are more and more complex (abstract). They create feature hierarchy (Fig. 7.3).

Because of their architecture, CNNs are well suited to image processing. They preserve spatial neighborhood in pixel processing. With the years CNNs have become deeper and deeper and bring in new concepts improving CNN's performance. The first CNN—LeNet composed only of a couple of layers, was developed to handwritten and machine-printed character recognition [118]. AlexNet [113]—the ILSVRC 2012 winner, is similar to LeNet but implements Dropout, ReLU activation function, and Max pooling. VGG [187] is a sequence of deeper networks trained progressively. GoogleNet [196] introduces inceptions modules and batch normalization [197]. ResNet [77] is a Deep Residual Learning, includes up to 152 layers a depth. It is about eight times deeper than VGG but still has lower complexity.

A typical application of CNN is classification [113, 227]. Another example of application is object detection in an image by so-called Region based Convolutional Networks (R-CNN) [57, 58, 171]. The network is able to draw bounding boxes over all of the objects on the input image. The process consists of two general components: the region proposal step and the classification step.

The next area of application is semantic image segmentation that needs to label each pixel with the class of its enclosing object or region. To solve this problem the Fully Convolutional (F-CNN) network was proposed [126]. The paper [154] proposes semantic segmentation algorithm by learning a deconvolution network. The approach from [11] consists of an encoder network, a corresponding decoder network followed by a pixel-wise classification layer. The architecture of the encoder network is similar to the 13 convolutional layers in the VGG16 network [187]. The decoder network maps the low-resolution encoder feature maps into full input resolution feature maps for pixel-wise classification.

Gong et al. [66] proposed a scheme, called multi-scale orderless pooling (MOP-CNN), for improving the invariance of CNN activations keeping the same their discriminative power. In 2015 Jaderberg et al. [100] presented the simple idea of Spatial Transformer Network. The network makes affine transformations to the input image to make the model more invariant to translation, scale, and rotation.

7.3.1.4 Recurrent Neural Networks

Unlike in feedforward neural networks, connections in Recurrent Neural Networks (RNN) [80] create directed cycles. In deep recurrent networks, very important role plays Long-Short Term Memory network (LSTM) [87]. The intuition behind this model is that humans do not throw everything away from the mind and start thinking from scratch every second, human thoughts have persistence. The recurrent neural networks contain loops that allow information to persist. LSTMs are a kind of recurrent neural networks, capable of learning long-term dependencies. They are suitable for large variety of problems, what allows their widely using. The architecture of LSTMs contains gates what prevents vanishing gradient problem. In standard recurrent networks, the repeating modules have a simple structure, e.g., single *tanh* layer. LSTMs have similar, chain-like structure but the repeating module has four interacting layers. The LSTM model introduces a new structure called a memory cell. A memory cell is composed of four main elements: an input gate, a neuron with a self-recurrent connection (a connection to itself), a forget gate and an output gate [67]. This structure enables to make decisions about what to store, and when to allow reads, writes, and erasures, by opening or closing gates. These gates are implemented with element-wise multiplication by sigmoids. They are differentiable, and therefore suitable for training by backpropagation. The deep LSTM network emerges by unfolding the network in time. Training is performed with Backpropagation Through Time (BPTT) [148].

Gated Recurrent Unit (GRU) is related to LSTM as both are utilizing different way of gating information to prevent vanishing gradient problem [28]. The GRU unit controls the flow of information like the LSTM unit, but without having to use a memory unit. It just exposes the full hidden content without any control. GRUs are relatively new. They are computationally more efficient (less complex structure) and have fewer parameters than LSTM. They are increasingly used. Because of deep recurrent neural networks ability to process pattern sequences, they are widely used in natural language processing [188, 194], speech recognition [68, 177].

7.3.2 Visual Attention Models Based on Deep Learning

Visual attention plays a role in the early stages of human vision; it provides a rich source of information useful in efficient conscious recognition. A model of attention can bring potential benefits in such applications as visual inspection in manufacturing processes, medical diagnosis, image and video analysis for different target tasks. Computer models of visual attention try to imitate the behavior of the human visual system. The main task of such model is to identify image regions that attract human attention. Attention in Neural Networks has a long history, particularly in application to image recognition [34, 116]. Saliency models predict the probability distribution of the location of the eye fixations over the image creating saliency map [95]. Section 7.2 shortly describes earlier methods to create saliency map. In this section, we shortly refer to those based on deep learning models. These models predict human eye fixations with strong semantic content. The examples of early approaches using DL are presented in [185, 204]. The common characteristic of the late methods is using Convolutional Neural Network (CNN) and end-to-end training. They differ in how they are using the pre-trained models of CNNs or use raw CNN models without pre-trained detectors [184], in the way of information concatenation, considering local or global features or the goal of the model use. In this group we can list the following studies: [36, 95, 217].

The *soft attention model* is a deterministic mechanism, fully differentiable, it can be plugged into some existing system, and propagates gradients through attentions mechanism at the same time they are propagated through the rest of the network. It takes as input all the hidden states and assigns relative importance of location i. In opposite, the *hard attention model* is a stochastic process. It samples a hidden state with the probabilities, the gradient being propagated is estimated by Monte Carlo sampling. This approach uses a probability of a given location i to choose the right place to focus on.

The example of hard attention is the model implemented by Google [145]. It consists of several parts, but the fundamental role plays LSTM network. Its current internal (hidden) state h_t contains information about previous state h_{t-1}. Based on the glimpse feature vector g_t produced by the glimpse network, and the previous state, the LSTM calculates the current state as $h_t = f_h(h_{t-1}, g_t; \theta_h)$ (θ_h are the parameters of the network). The location network and the action network use the internal state h_t to produce the next location l_t and the action l_a. The Gaussian distribution is used to generate next location. Because of this stochastic element, it is not possible to apply gradient learning rule, and the reinforcement learning is proposed. The extended version of this model, where the *Context network* is introduced, is presented in [9]. The soft attention is deterministic, so it is possible to learn the model end-to-end using standard backpropagation. Its popular application is image caption generation. As it is shown in Fig. 7.4, this model uses CNN to extract features of an image. Next, they are given as the input to the recurrent neural network.

Xu et al. [222] used LSTM that generates a caption for the image using an attention mechanism. It selectively focusses on parts of the image by weighting a subset of the features extracted by the convolutional neural network. Other advanced models of visual attention are described in many papers, e.g., [95, 157, 223, 233].

7.3.3 Embedding Semantic Features

A limited amount of labeled images collected in databases causes limitation of modern visual systems. Manually labeling images is the time-consuming and boring task, especially as the number of object categories grows. One possible remedy to this problem can be leveraging data from other sources. Such sources can be text data. Multi-modal models that combine robust visual features extracted from image data, and the linguistic features, extracted from linguistic data, produce better results

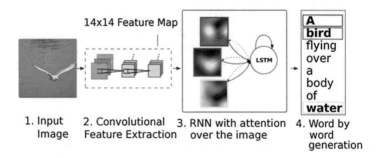

Fig. 7.4 The soft attention model [222]. Used with kind permission Kelvin Xu

comparing to uni-modal models. Recently deep visual-semantic embedding models are proposed and widely studied. They use labeled images, and the semantic information gleaned from the unannotated text for training to identify visual objects on images. As the linguistic features, extracted from text data, are used in image semantic features representations, in this section, we present the embedding semantic features models briefly.

To automatically process language it is necessary to solve two problems: how to encode word and how to build a language model. Language models try to compute the probability of a word w_t given its $n - 1$ previous words, i.e., $p(w_t|w_{t-1}, ..., w_{t-n+1})$. In the classical approach, we can calculate the probabilities of each word given its n previous words by applying the Markov chain rule. For a long time, the representation of word was the most simple: 1-of-n vector. It represents every word as a vector with all 0 s and one 1 at the index of given the word. Assuming that our vocabulary has only four words ($n = 4$): cat, dog, frog and tiger, the vector encoding the word *dog* is [0 1 0 0]. With such encoding, it is only possible to test equality between word vectors. That is why currently a distributed representation of a word is used. It ensures that semantic relationship between words is preserved in word vectors and implements dimensionality reduction. The new methods build a low-dimensional vector representation from the corpus of text, which preserves the contextual similarity of words.

Though the new methods of words encoding are not based on deep learning (they use shallow neural networks), they play the essential role in natural language processing delivering a new effective representation of text. In [140] two architectures for learning word embedding are presented. The methods are implemented in the form of feed-forward neural network that takes words from a vocabulary V as an input and embeds them as vectors of weights W_{1VxN} into a lower dimensional space N. The network is then fine-tuned by backpropagation. W_{1VxN} are word embedding that has compact structure and preserves semantic relationships between words (singular versus plural relation, gender, jobs). CBOW method takes as an input the context words of a given word w_t. Each word as an input is represented by a vector encoded by 1-of-$|V|$ rule ($|V|$ is the number of words in the vocabulary). The network output predicts the word w_t. The output layer is a softmax layer, so the word w_t is achieved as the one that satisfies $argmax_i\, p(w_i|w_{i+1}, ..., w_{i-n+1})$ of the network responses. A hypothetical word vector being embedding for the word *dog* obtained from the hidden layer weights could be [0.99 0.56 0.69 0.22 0.32 0.71 0.11]. The number of elements in this vector is equal to the number of hidden neurons in the network. Each element of this vector represents a latent feature. In comparison to the thousands of words in the corpus, the network creates a compact representation with the number of elements in the embedding vector much smaller than the size of the corpus.

The Skip-gram model uses the word w_t to predict the surrounding words. The model calculates the probability of the surrounding words w_{t+j} given w_t. This network uses hierarchical softmax—a binary tree to represent all words in the vocabulary. The words are placed in the leaves. From the root to each leaf there exists a unique path. This path is used for estimation of the probability of the word represented by the leaf. In [143] Mikolov et al. introduced a further improvement to the

Skip-Gram with Negative Sampling. The paper [60] explains how the skip-gram model with negative sampling works in detail. The negative-sampling is more efficient than Skip-gram model. It optimizes a different objective, related to the joint distribution of word and context. The words and contexts representations are learned jointly. The above described methods are offered as a tool that is known as Word2Vec. Another example is GloVe [161]. The paper [81] shows how to use the definitions found in everyday dictionaries to bridge the gap between lexical and phrasal semantics. The task of the neural language embedding models is mapping dictionary definitions (phrases) to (lexical) representations of the words defined by those definitions.

Another branch of research connected with word embedding is focused on recurrent neural networks [141]. Embedding the whole text into a sequence of vectors is a much more powerful way to make the word representations context-specific. Bidirectional RNNs are used for encoding the vectors into a sentence matrix. The rows of the matrix can be perceived as token vectors. They are sensitive to the sentential context of the token. In most cases, LSTM and GRU architectures are applied for this reason. The essential element used here is an attention mechanism. The new variation is presented by Parikh et al. in [158]. They introduced an attention mechanism that takes two sentence matrices and outputs a single vector. Yang et al. [223] insert an attention mechanism which takes a single matrix and gives a single vector as output. Instead of a context vector derived from some aspect of the input, they computed summary regarding a context vector learned as a parameter of the model.

Zhang et al. [231] show how to apply temporal convolutional networks to text understanding from character-level inputs all the way up to abstract text concepts. They show that temporal CNNs can achieve astonishing performance.

The work [224] compares CNN and RNN techniques and concludes that CNN is supposed to be good at extracting position invariant features and RNN at modeling units in sequence. The paper [188] introduces temporal hierarchies to the Neural Language Model (NLM) with the help of a Deep Gated Recurrent Neural Network with adaptive timescales to help represent multiple compositions of language.

7.4 Deep Learning in Bridging the Semantic Gap

Section 7.2 gave an insight how researchers tried to narrow the semantic gap in image retrieval before DL. In this section, we show what new possibilities arose with the existence of DL.

Considering the scale of image analysis, we distinguished three levels of bridging the semantic gap problem using DL: features, knowledge, and inference. They are ordered with growing complexity of image analysis needed. This corresponds to the results they offer.

The lowest level refers to features—an image representation that in the best way reflects the content of a given image. DL offers much in this area because it enables automatic acquisition of features during training.

The next level is building common sense knowledge by learning temporal and spatial knowledge from visual data aligned with textual data. It is a way for better understanding image content. Knowledge acquisition and then its integration in the form suitable for reasoning is another vital aspect in bridging the semantic gap where DL can be helpful.

The highest level is making an inference about a scene, event or objects in an image based on the second level. It is the most demanding process to bridge the semantic gap, and now it can also be supported by DL to some extent.

The feature level can be enriched in visual emotion analysis. It provides valuable semantic meanings about image content, which can hardly be represented by low-level visual features. You et al. [226] describe an attempt to collect a large dataset prepared to analyze and predict people's emotional reaction towards images. Affective analysis of images can be based on the texts surrounding images. Liu et al. [124] propose textual features to efficiently capture emotional semantics from the short text associated with images based on word similarity. Their approach combines visual and textual features and provides promising results for the visual affective classification task.

To establish the similarity/dissimilarity between images, we measure the feature distance, or another similarity function is calculated. DL makes an offer of efficient solutions in matching images without the necessity of using rigid measures. The following subsections present the role of DL in the mentioned above levels.

7.4.1 Feature Level

In classical approaches, features are manually encoded vectors of attributes, describing shared characteristics among image categories. In the process of deep model training, the features can be assigned automatically. Visual data is one of the most abundant sources of information, but unfortunately, in most of the cases, it is not adequately exploited due to the difficulty in analyzing this kind of data.

Visual feature extraction, described in Sect. 7.4.1.1 is the primary step in all image retrieval methods. Lately, deep models are used to generate images on the basis of the text. One can imagine that visual features extracted from models trained with these images express in a more powerful way the semantics included in the text. Section 7.4.1.2 presents this problem. In many cases, visual features are not sufficient to relevant image retrieval due to the difficulty in analyzing this kind of data. The solution lies in considering complementary information from other data modalities. Section 7.4.1.3 shows how to combine visual and textual features.

7.4.1.1 Visual Feature Extraction

All approaches described in this subsection use deep models to produce visual feature vector. As an input, they get raw images. The first attempts in the application of deep neural networks to extract visual features use autoencoders. In [112] authors obtained compact visual representation of visual features by learning Deep Belief Networks (DBNs) [85], consisting of stacked Restricted Boltzmann Machines (RBMs). They trained the network on 1.6 millions 32×32 color normalized images. In the first RBM, most of the hidden units learned to be high-frequency monotone Gabor-like filters, and most of the remaining units became lower frequency filters responding to color edges. The codes obtained by looking at the entire image are suitable for capturing global image structure, but they are not invariant to the image transformations. Therefore authors trained the network treating an image as a bag of patches. 256-bit code extracted from the fourth layer of the network gave qualitatively and quantitatively better results in comparison to other methods.

A convolutional neural network (CNN) has a leading position on feature extraction and representation for CBIRs. The paper [38] analyzes feature representation acquired from fully connected layers of CNN. Authors conduct experiments on both ImageNet-2012 and an industrial dataset provided by Sogou platform using AlexNet [113]. The results demonstrate that the features extracted from the first and the second fully connected layers of AlexNet perform the best on the datasets from unseen categories. Authors noticed that the performance for features obtained by CNN on in-class data and out-of-class data is not consistent. The higher layer level, the higher the difference.

In the research described in [123] authors contemplate the semantic gap between low-level features and high-level semantic features of the convolutional network. They consider that the high-level features extracted from higher levels of a deep model are more abstract. It means they better express semantic concepts. In their experimental study, they used features of the sixth fully connected layer—F6 from the adapted convolutional network—LeNet-L and eighth fully connected layer—F8 from AlexNet. In the experiments they show that combined features from both layers outperform the results obtained on the basis of features from F6 and F8, and the results with F8 were better than F6.

Summing up, research in this area apart from DBN network concentrates mainly on using various architectures of convolutional networks, studying an influence of network depth and the choice of the fully connected layer to extract features on the efficiency of image recognition. In this place, it is worth noticing the first work focusing on visualizing features in the subsequent layers of CNN [229]. Human design of visual feature is now the past. The main advantage of using DL to this task is automatization of the visual feature extraction process.

7.4.1.2   Visual Descriptors Acquired from Images Generated on the Basis of Text Descriptions

This subsection presents a potential solution to the problem of bridging a semantic gap that was not applied to image retrieval so far. The possibility of plausible images generation based on detailed text descriptions is a relatively new achievement in DL history. Translating caption describing a hypothetical image using Generative

Adversarial Network (GAN) [64] gives a variety of images corresponding to the description. GAN is implemented as two neural networks contesting with each other in a zero-sum game framework. The newly generated images well illustrate semantic concepts included in the text description. They can also recover some features. In this context paper [182] is worth citing here where GAN was able to improve features in astronautical images of galaxies beyond the deconvolution.

Training DL model using this variety of generated images and then extracting features should give features that represent semantic image content in an adequate way for retrieval purpose. The approach based on this concept would get as an input text description, then it will generate images corresponding to it and next using these images the process will end with feature extraction as described in Sect. 7.4.1.1 or in 7.4.1.3. This most certainly means that deep model trained on the generated images corresponding to the text descriptions would create more adequate descriptors reflecting the semantic content of the image so that it can be perceived as the future direction of research in this area.

At the current stage, research in DL is focused on the methods of generating images from a text description. Usually, as a language model, a standard bidirectional LSTM network is used, and the image model is a variational recurrent autoencoder with visual attention. Visual attention allows determining which image part should be generated in response to the current text part in the caption. The model proposed in [137] iteratively draws patches on canvas while attending to the relevant words in the description. The model extends the Deep Recurrent Attention Writer (DRAW) [70]. The images generated by the model are refined in a post-processing step by a deterministic Laplacian pyramid adversarial network, first presented in [35]. At each level of the pyramid, a separate generative convolutional network model is trained using GAN. All stages were conditioned on the same skip-thought vector. This idea was proposed to reconstruct the surrounding sentences of an encoded passage. Sentences that share semantic and syntactic properties are mapped to similar vector representations [110]. The authors of [35] claim that samples drawn from their model are of significantly higher quality than alternative approaches.

To sum up, research in this area focuses on the methods of synthesizing new, photorealistic images from a text. GAN is of a great potential, and many researchers are exploring the ideas behind it. As an example the work of Zhang et al. [234] can be mentioned, where the first GAN draws the object following primary colors and shape constraints from given text descriptions, and GAN on the higher level corrects the defects of the first GAN and adds more photo-realistic details. Another example is presented in [170]. It proposes a new model of GAN named GAWNN from Generative Adversarial What-Where Network that synthesizes images given instructions describing what content to draw in which location.

We can predict that in the future the leading role will play development of GAN architectures. Another research is the development of attention methods that make more natural to combine the recurrent network responsible for text processing with the network generating an image.

7.4.1.3 Combined Visual and Text Features

Finding images accompanied by unstructured related text description is common. This multimodal data originated from the same source tend to be correlated therefore multimodal semantic analysis helps overcome the visual semantic gap.

This approach simultaneously incorporates descriptions from an image and a text and finds common space for them. Finding it relies on projecting image features and text features referring to the image into a common space. It is sometimes called multi-modal embedding.

The research on multi-modal embedding based on deep learning is widely presented in the literature [10, 109, 128, 191, 203].

In [10] authors combine visual features from a convolutional network with information obtained from a multi-layer perceptron (MLP) and produce a set of linear output nodes. Wikipedia articles about a particular object are passed through MLP that produces some semantic features based on the text. This deep neural network model maps raw text and image pixels to a joint embedding space. It was used as Zero-Shot Learning model that learns to predict unseen image classes from encyclopedia articles.

Kiros et al. in [109] introduced two methods based on the log-bilinear model (it uses a feed-forward neural network with a single linear hidden layer) which operates on word representation vectors. The authors show how to learn word representations and image features together with jointly training language models and a convolutional network that processes an image. An image-text multimodal neural language model can be used to retrieve images given complex sentence queries, retrieve phrase descriptions given image queries, as well as generate text conditioned on images. The method outperformed a strong bag-of-words baseline for description and image retrieval.

In [191] the authors propose a Deep Boltzmann Machine (DBM) for learning a generative model of multimodal data. The model consists of image-specific two-layer DBM to model the distribution over real-valued image features and text-specific two-layer DBM to model its distribution over the word count vectors which are combined to model the joint distribution over image and text inputs. Authors show that the model can be used to create fused representations by combining features across modalities. By sampling from the conditional distributions over each data modality, it is possible to create these representations even when some data modalities are missing. The probability of generating sentences given the corresponding image can serve as the affinity metric for retrieval. They demonstrated that their multi-modal model helps classification and retrieval even when only unimodal data is available at test time.

An interesting approach is described in [128], where Ma et al. have proposed a multi-modal convolutional neural networks (m-CNNs) for matching image and sentence. The m-CNNs rely on two convolutional networks (image CNN and matching CNN) and multilayer perceptron (MLP) that takes the joint representation as input and produces the final matching score between image and sentence. The image CNN is used to generate the image representation for matching the fragments consisted of words of the sentence to compose different semantic fragments of the sentence. The matching CNN takes the encoded image representation and word representations as input and produces the joint representation.

Wang et al. [213] propose another method preserving structure, for learning joint embeddings of images and text. They use a simple model composing of two-branch neural network with multiple layers. Each branch consisted of fully connected layers with ReLU between them, followed by L2 normalization at the end. As the authors claim, the retrieval results on Flickr30K and MSCOCO datasets exceeded state of the art.

Mao et al. in [136] describe several CNN-RNN based multimodal models to learn useful word embeddings. In the experiments, they show that visual information significantly helps the training of word embeddings. In another work [135], authors experimentally confirmed that the model benefits from incorporating the visual information into the word embedding, and a weight sharing strategy is crucial for learning such multi-modal embeddings.

Paper [47] presents a new deep visual-semantic embedding model trained to identify visual objects using both labeled image data as well as semantic information aggregated from an unannotated text. They used a simple neural language model for learning semantically-meaningful, dense vector representations of words and CNN for visual object recognition. The last output layer is softmax. Authors construct a deep visual-semantic model by taking the lower layers of the pre-trained visual object recognition network and re-training them to predict the vector representation of the image label text as learned by the language model. The experiments show that this model can make correct predictions across thousands of previously unseen classes by leveraging semantic knowledge elicited only from unannotated text.

Collell et al. [31] present a method capable of generating multi-modal representations in a fast and straightforward way by using pre-trained unimodal text and visual representations as a starting point. GloVe [161] was used for word embeddings and two different CNN networks for visual embeddings. Authors found that neither vision nor language is superior to the other, but they instead dominate in various attribute types. Vision proves better at capturing form and surface, color and motion attributes while language proves better at encyclopedic and function attributes.

An obvious conclusion from all papers discussed in this subsection is that joint information (text and visual) is beneficial for image recognition and deep models and provides greater representational power than methods based on linear projections. In this context, it is worth adding the result from the research in [30] showing that visual and textual representations encode different semantic aspects of concepts. The authors claim that neither vision nor language is superior to the other in grasping every aspect of meaning. They dominate in different attribute types.

Summing up, the first deep models used in multi-modal embedding were: deep Boltzmann machines [191] and RBMs [151]. In the last years popular models are based on CNN [128] to process visual data and MLP or LSTMs [203] to embed text features.

In this subsection, we have assumed that text describing an image is available as a text associated with an image. In Sect. 7.4.2.1 we show how such annotating text can be automatically generated using deep learning approach.

7.4.1.4 Modelling Similarity Function

In image retrieval, the problem of similarity between images plays a significant role. To model similarity function, the paper [214] proposes Siamese network, which consists of two weight-sharing networks running on two input images in parallel. It is trained on pairs of images labeled as similar or dissimilar. This approach can be used with a contrastive loss to minimize the distance between related images (maximize the distance between dissimilar images) in the feature space. The siamese network approach has been extended to triplet networks, where an image is given into the network together with one similar and one dissimilar image during each training step [62, 88].

A similar approach that uses the siamese network for similarity evaluation presents the paper [167]. Authors use predefined features from Convolutional Neural Network. The proposed architecture learns a new distance feedback.

7.4.2 Knowledge Level

In this section, we focus on knowledge that can be acquired from an image. Knowledge is an understanding of an image expressed by text descriptions. It is obtained automatically through a learning process. Another aspect of knowledge is its formal representation that can be used to reasoning about new concepts or a relation between objects in an image.

7.4.2.1 Semantic Image Interpretation and Annotation

The semantic image interpretation is a hard problem for computers because of many reasons: the sensory gap that results from a projection of reality to 2D representation, semantic gap and scaling gap that is understood as a balance between expressivity/complexity and scaling of models [189, 190]. Semantic image interpretation/annotation relies on the automatic generation of text/labels describing an image. As an input, the method takes a raw image. On this basis it produces labels or text descriptions in the form of captions.

Annotation of an image is more easy with image segmentation. There are many approaches based on deep models [126]. The readers interested in this topic are directed to a comprehensive survey that can be found in [54].

The model shown in [104] takes a dataset of images and their sentence descriptions and learns to associate their fragments. In images, fragments correspond to object detections and scene context. In sentences, fragments consist of relations. The task was to retrieve relevant images given a sentence query, and conversely, relevant sentences given an image query. The model breaks down both images and sentences into fragments and reasons about their alignment. To detect objects as image fragments, a special CNN is applied, and sentence fragments are identified by sentence dependency tree relations.

Another interesting approach is described in [24]. Authors use deep recurrent neural network to automatically generate captions for images. The proposed RNN model is bi-directional. This means that the network can generate image features from sentences and sentences from image features. This is very helpful in the content based image retrieval, because a query can be expressed in dual form: as a description or as an image. In the case of an image description generation, the LSTM network generates a probability of the word w_t at time t as the next word in the sequence of previously generated words $W_t = w_1, w_2, ...w_{t-1}$ and visual features V. The built model enables to compute the likelihood of the visual features V given a set of words W_t for generating visual features or for performing image search.

Paper [134] also describes multi-modal recurrent neural network generating description of the image but it interacts with convolutional network that extracts visual features from an image.

Authors of [27, 237] describe a novel image annotation framework which explores a unified two-stage learning scheme by learning to fine-tune the parameters of deep neural network with respect to each individual modality, and by learning how to find the optimal combination of diverse modalities simultaneously. Another example of image captioning is described in [208]. It presents the system that can automatically view an image and generate a description in English. The system is based on a convolutional neural network that encodes an image, followed by a recurrent neural network that generates a corresponding sentence.

In [46] authors detect words by applying a CNN to image regions and integrating the information with Multiple Instance Learning. Paper [37] describes recurrent convolutional architectures which is end-to-end trainable and suitable for large-scale visual understanding tasks which can be applied for activity recognition, image captioning, and video description. Relatively new approach in automatic image captioning is using visual attention [222, 228].

Summing up, popular deep models applied in image description are CNN to process image and RNN to process text or image. Visual attention models are more and more applied. The automatically extracted text describing an image can be used to build multi-modal image embeddings, as it was mentioned in Sect. 7.4.1.3 or it can be further processed to extract new knowledge about an image by utilising natural language processing methods.

7.4.2.2 Formal Knowledge

The formal knowledge representation facilitates knowledge reusing and sharing in a machine-processable way. Wang et al. in [38] present a survey of existing ontology-based approaches and other formal representation based data mining algorithms. In another paper [212], Wang illustrates how learning models with deeper architectures are capable of constructing better data representations for machine learning tasks and information retrieval. In this paper, the author defines the term semantic deep learning, as a combination of deep learning techniques and formal knowledge representation. The paper focuses on ontologies because they are the most popular formalism to specify domain semantics. Using it, we can also reduce the semantic gap by annotating the data with rich semantics. There are also examples of other knowledge representation usage, as Formal Logic Description (FCA) [8] to image interpretation.

In [213] authors build an ontology based on deep learning model. It formally encodes the concepts and relations in the domain of the data label. Their deep ontology model is based on the RBM network. The primary network learns the first concept. Then, the architecture is extended with subconcepts. The model is trained in the unsupervised and next in a supervised way. Another example of automatic building an ontology is described in [164]. In this paper, authors trained recurrent neural network architectures to extract OWL formulae from a text. Nguyen et al. in [152] explain how knowledge base can be integrated with the representation learning using a deep network Deep Semantic Structured Model (DSSM), either through an enhanced knowledge-based representation of the document and the query or as a translation representation bridging the semantic gap between the document and the query vocabulary.

It is evident that formal knowledge can reduce the semantic gap. Its effective application includes: building a dedicated visual concept ontology as an intermediate level between image features and application domain concepts: [12, 132, 139, 166, 200] and using concrete domains to link high level concepts to their specific representations into the image domain [97], where each application domain concept is linked to its representation in the image domain. Deep ontology models that can learn relationships between concepts using data can be widely used in semantic data annotation [44, 49, 156], semantic aware preprocessing [162] and producing semantic rich data mining results [220].

The benefit that can be delivered by using formal knowledge representation is integration capability with another source of information and possibility to produce a high-level specific knowledge. It also enables inferring about an image.

7.4.3 Inference Level

Inference about an image is the most demanding process for computers. It uses knowledge and its formalization. It allows to reason new knowledge and build amazing automatic systems answering questions about an image. For answering questions, the system must perform automatic reasoning because this knowledge is not explicitly given in the image text description. Deep models are also helpful in this task.

Lippi in [121] addresses challenges that relate to deep learning application in symbolic reasoning. This work emphases the lack of explanation of shallow and deep neural networks. Researchers realize that a combination of neural networks and symbolic reasoning could be beneficial. Garcez et al. [52] discuss present achievements and critical challenges for neural-symbolic integration.

In the current stage of DL, rather than logic-based formal reasoning researchers explore to perform relational reasoning about images using deep models. Santaro et al. [179] explore the ability for deep neural networks to perform complicated relational reasoning with unstructured data. They describe how to use Relation Networks (RNs) as a simple plug-and-play module to solve problems that focus on relational reasoning.

Hohenecker et al. [90] propose a new model for statistical relational learning that is built upon deep recursive neural networks, and give experimental evidence that it can easily compete with, or even outperform, existing logic-based reasoners on the task of ontology reasoning.

Malinowski et al. [133] propose the system Ask Your Neurons, answering the questions about an image. The system consists of CNN processing images and LSTM operating on questions and producing answers. It also uses visual attention. A multimodal fusion module combines visual and text vector spaces into another vector based on which the answer is decoded. Authors perform an analysis of the large-scale dataset showing competitive performance.

Jonson et al. [103] propose a model for visual reasoning. It consists of a program generator that constructs an explicit representation of the reasoning process to be performed, and an execution engine that executes the resulting program to produce an answer. Few other examples of systems answering the questions about images are [4, 129, 155]. A novel visual attention model [180] is worth noticing.

To achieve a success, models need to understand the interactions and relationships between objects in an image. Computers need to identify the objects in an image and the relationships between them. There is a big need for new generation of benchmarking datasets [111] including the images densely annotated with numerous region descriptions, objects, attributes, and relationships (Fig. 7.5).

A scene graph represents all the objects and relations in a scene. The scene graph representation has been shown to improve semantic image retrieval [102, 183]. The Genome dataset [111] contains more than 108 K images and can be used to develop deep models that enable automatic reasoning about the scene in the image. It is crucial in applications, such as image search, question answering, and robotic interactions. Other benchmarking datasets for useful cognitive scene understanding and reasoning tasks are MS-COCO [120] and VQA [5]. These datasets can help train models that can learn from contextual information from multiple images.

7.5   Conclusions

Deep Learning has enabled a huge jump solving many problems, including *Content Based Image Retrieval*. Many ideas in CBIR were initiated before DL arose. It has revolutionized the automatic visual features extraction, creating visual features based on the text describing an image, multi-modal features embedding and modeling similarity function between images. Figure 7.6 contains a table that summarizes how DL bridges the semantic gap in image retrieval. Of course, we have to mention that this table does not contain all research on this subject, only some exemplary literature is written in the *References* column.

Results of DL in automatic image annotation and captioning are very impressive. Supporting knowledge representation by DL is also worth emphasizing. Research is developed for relational reasoning about images that enables building systems answering questions referring to the visual content of an image.

Up to now, none other approach has given so good integration of vision and text that naturally expresses semantics of an image.

Some issues described here influence the bridging the semantic gap indirectly. As an example, one can indicate a new word embedding technique that enables high progress in natural language processing and visual attention models, which is very important in image captioning and natural language processing.

The DL results heavily depend on large amounts of labeled data. That is why there is a need for new datasets that could enable the learning process. Another desirable solution is a development of novel deep unsupervised learning methods. Generating labeled corpora is very complex, time-consuming and costly operation, whereas unsupervised data are everywhere.

Contextual adaptation allows avoiding the necessity of huge dataset. It implies adapting behavior depending on context. Bartunov et al. in [13] described a new class of deep generative model called generative matching network (GMNs) which

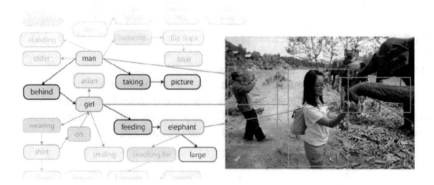

Fig. 7.5   Representation of an image in Genome dataset [111]

	Level	Characteristics	Models	References
Feature	Visul decriptors	• Features automatically assigned • The higher layer of NN, the more complex features • Learning transfer is popular	• RBM • CNN	[112], [211], [123], [229]
	Visual descriptor generation based on text	• Recurrent network as a language model • Variational recurrent autoencoder as an image model • Visual attention useful	• LSTM • VAE • GAN	[137], [110], [35], [167]
	Joint visual and semantic features	• Multi-modal descriptors (semantic and visual) • Visual-semantic embedding for bidirectional descriptor generation • Wealth of deep models presented in literature	• LSTM • CNN • MLP • DBM • Autoencoder	[169], [10], [109], [191], [203], [128], [151], [47], [31], [136]
	Similarity Function	• Model based on DL • Siamese network • The model can be extended to triplet network	• CNN	[214], [88], [62], [167]
Knowledge	Automatic image annotation and captioning	• Automatic annotation easier with image segmentation • Image captioning made by association part of text and an image fragment • Application of visual attention	• LSTM • bidirectional RNN • CNN	[189], [190], [54], [104], [24], [237], [27], [61], [44], [49], [156]
	Knowledge representation	• Automatic building formal knowledge representation • Formal knowledge - ontologies, Formal Logic Description • Visual attention useful	• RBM • RNN	[212], [8], [213], [164], [152], [200], [12], [132], [166], [139], [97]
Inference		• Inference mainly about relation of the image objects • Question answering systems are more and more popular in research • Interactions and relationships between objects in an image are important • Visual attention systems are applied • Objects and relations are represented as a scene graph • Benchmark datasets: Genome, MS-COCO and VQA • Symbolic reasoning by DL is currently not possible	• LSTM • RNN • CNN	[52], [179], [90], [133], [103], [129], [155], [4], [111],

Fig. 7.6 Summary table of DL application to bridge the semantic gap in image retrieval

is inspired by the recently proposed matching networks for one-shot learning in discriminative tasks.

Encoding human intention or emotions triggered by images is a big challenge. There are some initial attempts [106] using DL models to recognize them. Recognition of emotion triggered by an image also will contribute to bridging the semantic gap in image retrieval.

Modern deep learning methods have made tremendous progress solving many problems referring to the semantic gap, but it is unlikely they will solve all of them. Although there is some initial research in this area [94], the explanation of how DL models perform reasoning is still a challenge.

Expectations that deep networks will adequately address the task of symbolic reasoning with deep learning is also still an issue. The solution may lie in a new class of neural network architectures, such as Neural Turing Machines [69], Memory Networks [219], Neural Reasoner [160] that combine inference with long-term memories. A new idea—the Neural Theorem Prover (NTP) follows neural-symbolic approaches to automated knowledge base inference, but they are differentiable concerning representations of symbols in a knowledge base and can thus learn representations of predicates, constants, as well as rules of predefined structure.

The popularity of deep learning paradigm and the use of DL in image analysis is growing intensively. The number of publications listed in a reputable scientific research database—Web of Science, is the evidence of it. In recent years there has been a great increase in a number of scientific publications in the field of deep learning in connection with image analysis, image retrieval and semantic image analysis. Figure 7.7 presents the graph of a number of publications (in particular topics) in

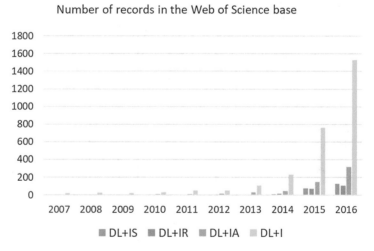

Fig. 7.7 Numbers of publications returned by the Web of Science database as answers to the queries containing following topics: DL+I—deep learning image; DL+IA—deep learning image analysis; DL+IR—deep learning image retrieval; DL+IS—deep learning image semantic

last ten years. Considering the Web of Science base and the period of last five years, the number of publication on Deep Learning topic increased from 1076 in 2013 to 4588 in 2017. At the same time, the number of publications on topic Deep Learning in conjunction with Image Analysis increased from 47 to 318; Image Retrieving— from 17 to 106; Image semantic—from 5 to 130. These data demonstrate the growing interest in the research area covered by the present chapter.

References

1. Agostinelli, F., Hoffman, M., Sadowski, P., Baldi, P.: Learning activation functions to improve deep neural networks (2014). https://arXiv.org/abs/1412.6830
2. Akcay, S., Kundegorski, M.E., Devereux, M., Breckon, T. P.: Transfer learning using convolutional networks for object recognition within X-ray Baggage security imaginary. In: International Conference on Image Processing ICIP, IEEE, pp. 1057–1061 (2016)
3. Alzubi, A., Amira, A., Ramzan, N.: Semantic content-based image retrieval: a comprehensive study. J. Vis. Commun. Image Representation **32**, 20–54 (2015)
4. Andreas, J., Rohrbach, M., Darrell, T., Klein, D.: Learning to compose neural networks for question answering. In: NAACL (2016). https://arXiv.org/abs/1601.01705
5. Antol, S., Agrawal, A., Lu, J., Mitchell, M., Batra, D., Zitnick, C.L., Batra, D., Parikh, D.: VQA: visual question answering. In: Proceedings of International Conference on Computer Vision (ICCV), pp. 4–31 (2015)
6. Arandjelovic, R., Zisserman, A.: Three things everyone should know to improve object retrieval. In: Proceedings of the 2012 IEEE Conference on Computer Vision and Pattern Recognition (CVPR). IEEE Computer Society, Washington, DC, USA, pp. 2911–2918 (2012)
7. Arik, S.O., Chrzanowski, M., Coates, A., Diamos, G., Gibiansky, A., Kang, Y., Li, X., Miller, J., Ng, A., Raiman, J., Sengupta, S., Shoeybi, M.: Deep voice: real-time neural text-to-speech. In: ICML (2017). https://arXiv.org/pdf/1702.07825.pdf
8. Atif, J., Hudelot, C., Bloch, I.: Explanatory reasoning for image understanding using formal concept analysis and description logics. IEEE Trans. Syst. Man Cybernetics Syst. **44**(5), 552–570 (2014)
9. Ba, J., Mnih, V., Kavukcuoglu, K.: Multiple object recognition with visual attention (2014). https://arXiv.org/abs/1412.7755
10. Ba, J. Swersky, K. Fidler, S. Salakhutdinov, R.: Predicting deep zero-shot convolutional neural networks using textual descriptions. In: Proceedings of the 2015 IEEE International Conference on Computer Vision (ICCV), pp. 4247–4255 (2015)
11. Badrinarayanan, V., Kendall, A., Cipolla, R.: SegNet: a deep convolutional encoder-decoder architecture for image segmentation (2016). https://arXiv.org/pdf/1511.00561.pdf
12. Bagdanov, A.D., Bertini, M., Del Bimbo, A., Serra, G., Torniai, C.: Semantic annotation and retrieval of video events using multimedia ontologies. In: International Conference on Semantic Computing (ICSC07), pp. 713–720 (2007)
13. Bartunov, S., Vetrov, D.P.: Fast adaptation in generative models with generative matching networks. In: ICLR 2017 (2017). https://openreview.net/pdf?id=r1IvyjVYl
14. Bay, H., Tuytelaars, T., Van Gool, L.: SURF: speeded up robust features. In: European Conference on Computer Vision (ECCV), pp. 404–417. Springer, Heidelberg (2006)
15. Bengio, Y.: Practical recommendations for gradient-based training of deep architectures (2012). https://arXiv.org/abs/1206.5533
16. Bengio, Y. Boulanger-Lewandowski, N. Pascanu, R.: Advances in optimizing recurrent networks (2012). http://arXiv.org/abs/1212.0901
17. Blei, D.M., Ng, A.Y., Jordan, M.I.: Latent Dirichlet allocation. J. Mach. Learn. Res. **3**, 993–1022 (2003)

18. Bojarski, M., Del Testa, D., Dworakowski, D., Firner, B., Flepp, B., Goyal, P., Jackel, L.D., Monfort, M., Muller, U., Zhang, J., Zhang, X., Zhao, J., Zieba, K.: End to end learning for self-driving cars. In: CVPR Proceedings (2016). https://arXiv.org/abs/1604.07316
19. Borji, A., Sihite, D.N., Itti, L.: Quantitative analysis of human-model agreement in visual saliency modeling: a comparative study. IEEE Trans. Image Process. **22**(1), 55–69 (2013)
20. Borji, A., Cheng, M.-M., Jiang, H., Li, J.: Salient object detection: a benchmark. IEEE Trans. Image Process. **24**(12), 5706–5722 (2015)
21. Bruce, N., Tsotsos, J.: Saliency based on information maximization. NIPS **06**, 155–162 (2006)
22. Bruce, N., Wloka, C., Frosst, N., Rahman, S., Tsotsos, J.: On computational modeling of visual saliency: examining whats right, and whats left. In: Vision Research, vol. 116, Part B, pp. 95–112 (2015)
23. Cao, Z., Simon, T., Wei, S.E., Sheikh, Y.: Realtime multi-person 2D pose estimation using part affinity fields. In: CVPR Proceedings (2017). https://arXiv.org/abs/1611.08050
24. Chen, X., Zitnick, C.L.: Mind's eye: a recurrent visual representation for image caption generation. In: IEEE Conference on Computer Vision and Pattern Recognition (CVPR), pp. 2422–2431 (2015)
25. Chen, D., Yuan, L., Liao, J., Yu, N., Hua, G.: StyleBank: an explicit representation for neural image style transfer. In: CVPR Proceedings (2017). https://arXiv.org/abs/1703.09210
26. Cheng, M.-M., Mitra, N.J., Huang, X., Torr, P.H.S., Hu, S.-M.: Global contrast based salient region detection. IEEE Trans. Pattern Anal. Mach. Intell. **37**(3), 569–582 (2015)
27. Chengjian, S., Zhu, S., Shi, Z.: Image annotation via deep neural network. In: 14th IAPR International Conference on Machine Vision Applications (MVA) (2015)
28. Chung, J., Gulcehre, C., Cho, K., Bengio, Y.: Empirical evaluation of gated recurrent neural networks on sequence modeling (2014). https://arXiv.org/pdf/1412.3555v1.pdf
29. Clevert, D.-A., Unterthiner, T., Hochreiter, S.: Fast and accurate deep network learning by exponential linear units (ELUs) (2015). https://arXiv.org/abs/1511.07289
30. Collell, G., Moens, M.-F.: Is an image worth more than a thousand words? On the fine-grain semantic differences between visual and linguistic representations. In: NIPS 2016 (2016). http://www.aclweb.org/anthology/C/C16/C16-1264.pdf
31. Collell, G., Zhang, T., Moens, M.-F.: Imagined visual representations as multimodal embeddings. In: International Conference on Computational Linguistics (COLING) (2017). https://aaai.org/ocs/index.php/AAAI/AAAI17/paper/view/14811/14042
32. Colombo, F., Muscinelli, S.P., Seeholzer, A., Brea, J., Gerstner, W.: algorithmic composition of melodies with deep recurrent neural networks. In: Proceedings of 1st Conference on Computer Simulation of Musical Creativity (2016)
33. Datta, R., Joshi, D., Li, J., Wang, J.Z.: Image retrieval: ideas, influences, and trends of the new age. ACM Comput. Surv. **40**(2), Article 5 (2008)
34. Denil, M., Bazzani, L., Larochelle, H., de Freitas, N.: Learning where to attend with deep architectures for image tracking. In: NIPS 2011 (2011). https://arXiv.org/abs/1109.3737
35. Denton, E.L., Chintala, S., Szlam, A., Fergus, R.: Deep generative image models using a Laplacian pyramid of adversarial networks. In: NIPS Proceedings (2015)
36. Dodge, S., Karam, L.: Visual saliency prediction using a mixture of deep neural networks (2017). https://arXiv.org/pdf/1702.00372.pdf
37. Donahue, J., Hendricks, L.A., Rohrbach, M., Venugopalan, S., Guadarrama, S., Saenko, K., Darrell, T.: Long-term recurrent convolutional networks for visual recognition and description (2016). https://arXiv.org/pdf/1411.4389.pdf
38. Dou, D., Wang, H., Liu, H.: Semantic data mining: a survey of ontology-based approaches. In: Proceedings of the 2015 IEEE 9th International Conference on Semantic Computing (IEEE ICSC 2015), pp. 244–251 (2015)
39. Dozat, T.: Incorporating Nesterov Momentum into Adam. In: ICLR Workshop (2016). http://cs229.stanford.edu/proj2015/054_report.pdf
40. Eakins, J.P. Graham, M.E.: Content-based image retrieval, JISC Technology Applications Programme Report 39. (1999). http://www.unn.ac.uk/iidr/CBIR/report.html
41. Eakins, J.P.: Towards intelligent image retrieval. Pattern Recogn. **35**, 3–14 (2002)

42. Enser, P., Sandom, Ch.: Towards a comprehensive survey of the semantic gap in visual image retrieval. In: Proceedings of the Second International Conference on Image and Video Retrieval (CIVR), pp. 291–299 (2003)
43. Eidenberger, H., Breiteneder, C.: Semantic feature layers in content based image retrieval: implementation of human world features. In: 7th International Conference on Control, Automation, Robotics and Vision, ICARCV 2002 (2002)
44. Erdmann, M., Maedche, A., Schnurr, H.P., Staab, S.: From manual to semi-automatic semantic annotation: about ontology-based text annotation tools. In: Proceedings of the COLING-2000 Workshop on Semantic Annotation and Intelligent Content, pp. 79–85 (2000)
45. Escalante, H.J., Hernadez, C.A., Sucar, L.E., Montes, M.: Late fusion of heterogeneous methods for multimedia image retrieval. In: Proceedings of the 1st ACM International Conference on Multimedia Information Retrieval, pp. 172–179 (2008)
46. Fang, H., Gupta, S., Iandola, F., Srivastava, R., Deng, L., Dollar, P., Gao, J., He, X., Mitchell, M., Platt, J.C., Zitnick, C.L., Zweig, G.: From captions to visual concepts and back (2015). https://arXiv.org/pdf/1411.4952.pdf
47. Frome, A., Corrado, G., Shlens, J., Bengio, S., Dean, J., Ranzato, M., Mikolov, T.: DeViSE: a deep visual-semantic embedding model. In: Annual Conference on Neural Information Processing Systems (NIPS) (2013)
48. Farabet, C., Couprie, C., Najman, L., LeCun, Y.: Learning hierarchical features for scene labeling. IEEE Trans. on Pattern Analy. Mach. Intell. **35**:8, 1915–1929 (2013)
49. Fu, J., Mei, T., Yang, K., Lu, H., Rui, Y.: Tagging personal photos with transfer deep learning. In Proceedings of International World Wide Web Conference (IW3C2) (2015)
50. Fukushima, K.: Neocognitron: a self-organizing neural network model for a mechanism of pattern recognition unaffected by shift in position. Biol. Cybern. **36**, 193–202 (1980)
51. Ganin, Y., Kononenko, D., Sungatullina, D., Lempitsky, V.: DeepWarp: photorealistic image resynthesis for gaze manipulation. In: 14th Proceedings of European Conference on Computer Vision – ECCV 2016, Amsterdam, The Netherlands, October 11–14, 2016, Part II, pp. 311–326 (2016)
52. Garcez, A.A., Besold, T.B., de Raedt, L., Foeldiak, P., Hitzler, P., Icard, T., Kuehnberger, K.-U., Lamb, L.C., Miikkulainen, R., Silver, D.L.: Neural-symbolic learning and reasoning: contributions and challenges. In: Proceedings of the AAAI Spring Symposium on Knowledge Representation and Reasoning: Integrating Symbolic and Neural Approaches (2015). https://aaai.org/ocs/index.php/SSS/SSS15/paper/view/10281/10029
53. Garcia-Diaz, A., Leboran, V., Fdez-Vidal, X.R., Pardo, X.M.: On the relationship between optical variability, visual saliency, and eye fixations: a computational approach. J. Vis. **12**(6), 17–17 (2012)
54. Garcia-Garcia, A., Orts-Escolano, S., Oprea, S.O., Villena-Martinez, V., Garcia-Rodriguez, J.: A review on deep learning techniques applied to semantic segmentation. arXiv:1704.06857v1. [cs.CV] 22 Apr 2017
55. Gatys, L.A., Ecker, A.S., Bethge, M.: Image style transfer using convolutional neural networks. In: 2016 IEEE Conference on Computer Vision and Pattern Recognition (CVPR), Las Vegas, NV, pp. 2414–2423 (2016). https://doi.org/10.1109/CVPR.2016.265
56. Gehring, J., Auli, M., Grangier, D., Yarats, D., Dauphin, Y.N.: Convolutional sequence to sequence learning (2017). https://arXiv.org/pdf/1705.03122.pdf
57. Girshick, R., Donahu, J., Darrell, T., Malik, J.: Rich feature hierarchies for accurate object detection and semantic segmentation. Technical Report (2013). https://arXiv.org/pdf/1311.2524v5.pdf
58. Girshick, R.: Fast R-CNN (2015). https://arXiv.org/pdf/1504.08083.pdf
59. Godfrey, L.B,. Gashler, M.S.: A continuum among logarithmic, linear, and exponential functions, and its potential to improve generalization in neural networks. In: Proceedings of 7th International Joint Conference on Knowledge Discovery, Knowledge Engineering and Knowledge Management, pp. 481–486 (2016). https://arXiv.org/abs/1602.01321
60. Goldberg, Y., Levy, O.: Word2vec explained: deriving Mikolov et al. Negative-sampling word-embedding method. (2014). https://arXiv.org/pdf/1402.3722v1.pdf

61. Gong, Y., Jia, Y., Leung, T., Toshev, A., Ioffe, S.: Deep convolutional ranking for multilabel image annotation (2015). https://arXiv.org/abs/1312.4894
62. Gordo, A., Almazn, J., Revaud, J., Larlus, D.: Deep image retrieval: learning global representations for image search. In: Proceedings of the European Conference on Computer Vision (ECCV), pp. 241–257. Springer, Cham (2016)
63. Goodfellow, I.J., Bulatov, Y., Ibarz, J., Arnoud, S., Shet, V.: Multi-digit number recognition from street view imagery using deep convolutional neural networks (2013). https://arXiv.org/pdf/1312.6082.pdf
64. Goodfellow, I., Pouget-Abadie, J., Mirza, M., Xu, B., Warde-Farley, D., Ozair, S., Courville, A., Bengio, J.: Generative adversarial networks (2014). https://arXiv.org/pdf/1406.2661.pdf
65. Goodfellow, I., Bengio, Y., Courville, A.: Deep Learning. MIT (2016)
66. Gong, Y., Wang, L., Guo, R., Lazebnik, S.: Multi-scale orderless pooling of deep convolutional activation features(2014). https://arXiv.org/abs/1403.1840
67. Graves, A.: Supervised sequence labelling with recurrent neural networks. Studies in Computational Intelligence, vol. 385, pp. 1–131. Springer, Heidelberg (2012)
68. Graves, A., Mohamed, A., Hinton, G.: Speech recognition with deep recurrent neural networks. IEEE International Conference on Acoustics, Speech and Signal Processing (ICASSP) (2013)
69. Graves, A., Wayne, G., Danihelka, I.: Neural turing machines (2014). https://arXiv.org/pdf/1410.5401.pdf
70. Gregor, K., Danihelka, I., Graves, A., Wierstra, D.: DRAW: A recurrent neural network for image generation. In: Proceedings of International Conference on Machine Learning ICML (2015)
71. Hare, J.S., Lewis, P.H., Enser, P.G.B., Sandom, C.J.: Mind the gap: another look at the problem of the semantic gap in image retrieval. In: Proceedings of Multimedia Content Analysis, Management, and Retrieval, vol. 6073 (2006)
72. Hare, J.S., Lewis, P.H.: Semantic retrieval and automatic annotation: linear transformations, correlation and semantic spaces. In: Imaging and Printing in a Web 2.0. World and Multimedia Content Access: Algorithms and Systems IV, pp. 17–21. (2010)
73. Harris, C.G., Pike, J.M.: 3D positional integration from image sequences. Image Vis. Comput. 6(2): 8790 (1988)
74. Haykin, S.: Neural Networks: A Comprehensive Foundation 2 edn. Prentice Hall (1998)
75. He, R., Xiong, N., Yang, L.T., Park, J.H.: Using multi-modal semantic association rules to fuse keywords and visual features automatically for web image retrieval. In: Information Fusion, vol. 12(3) (2010)
76. He, K., Zhang, X., Ren, S., Sun, J.: Delving deep into rectifiers: surpassing human-level performance on ImageNet classification. In: 2015 IEEE International Conference on Computer Vision, IEEE Computing Society, pp. 1026–1034 (2015). https://doi.org/10.1109/ICCV.2015.123
77. He, K., Zhang, X., Ren, S., Sun, J.: Deep Residual Learning for Image Recognition. In: CVPR (2016). https://arXiv.org/abs/1512.03385
78. Hein, A.M.: Identification and bridging of semantic gaps in the context of multi-domain engineering, abstracts of the 2010 Forum on Philosophy, Engineering and Technology. Colorado (2010). http://philengtech.org/wp-content/uploads/2010/05/fPET2010-abstracts-5-1-2010.pdf. Accessed on 16 Aug 2017
79. Hermann, K.M., Kocisky, T., Grefenstette, E., Espeholt, L., Kay, W., Suleyman, M., Blunsom, P.: Teaching machines to read and comprehend (2015). https://arXiv.org/pdf/1506.03340.pdf
80. Hermans, M., Schrauwen, B.: Training and analyzing deep recurrent neural networks. In: NIPS 2013 (2013). https://papers.nips.cc/paper/5166-training-and-analysing-deep-recurrent-neural-networks.pdf
81. Hill, F., Cho, K., Korhonen, A., Bengio, Y.: Learning to understand phrases by embedding the dictionary. Trans. Association Comput. Linguist. 4, 17–30 (2016)
82. Hinton, G., Salakhutdinov, R.: Reducing the dimensionality of data with neural networks. Science 313(5786), 504–507 (2006)

83. Hinton, G.E., Osindero, S., Teh, Y.W.: A fast learning algorithm for deep belief nets. Neural Comput. **18**(7), 1527–1554 (2006)
84. Hinton, G.E.: Learning multiple layers of representation. Trends Cognitive Sci. **11**, 428–434 (2007)
85. Hinton, G. E.: A practical guide to training restricted Boltzmann machines. Technical Report UTML2010-003. University of Toronto (2010)
86. Hinton, G., Deng, L., Yu, D., Dahl, G.E., Mohamed, A., Jaitly, N.: Deep neural networks for acoustic modeling in speech recognition: the shared views of four research groups. IEEE Sig. Process. Mag. **29**(6), 82–97 (2012)
87. Hochreiter, S., Schmidhuber, J.: Long short-term memory. Neural Comput. **9**(8), 1735–1780 (1997)
88. Hoffer, E., Ailon, N.: Deep metric learning using triplet network. In: International Workshop on Similarity-Based Pattern Recognition, pp. 84–92. Springer, Cham (2015)
89. Hofmann, T.: Unsupervised learning by probabilistic latent semantic analysis. Mach. Learn. **42**(1–2), 177–196 (2001)
90. Hohenecker, P., Lukasiewicz, T.: Deep learning for ontology reasoning (2017). https://arXiv. org/abs/1705.10342
91. Holder, C.J., Toby, P., Breckon, T.B., Wei, X.: From on-road to off: transfer learning within a deep convolutional neural network for segmentation and classification of off-road scenes. In: European Conference on Computer Vision, pp. 149–162. Springer, Cham (2016)
92. Hou, X., Zhang, L.: Dynamic visual attention: searching for coding length increments. In: NIPS08, pp. 681–688 (2008)
93. Hou, J., Zhang, D., Chen, Z., Jiang, L., Zhang, H., Qin, X.: Web image search by automatic image annotation and translation. In: 17th International Conference on Systems, Signals and Image Processing (2010)
94. Hu, Z., Ma, X., Liu, Z.: Harnessing deep neural networks with logic rules (2016). https:// arXiv.org/pdf/1603.06318.pdf
95. Huang, X., Shen, C., Boix, X., Zhao, Q.: SALICON: reducing the semantic gap in saliency prediction by adapting deep neural networks. In: Proceedings of 2015 IEEE International Conference on Computer Vision, ICCV 2015, vol. 11–18, December 2015, pp. 262–270 (2015)
96. Huang, A., Wu, R.: Deep learning for music (2016). arXiv:1606.04930v1 [cs.LG]. https:// cs224d.stanford.edu/reports/allenh.pdf
97. Hudelot, C., Atif, J., Bloch, I.: ALC(F): a new description logic for spatial reasoning in images. ECCV Workshops **2**, 370–384 (2014)
98. Ioffe, S., Szegedy, C.: Batch normalization: accelerating deep network training by reducing internal covariate shift (2015). https://arXiv.org/abs/1502.03167
99. Isola, P., Zhu, J.Y., Zhou, T., Efros, A.A.: Image-to-image translation with conditional adversarial networks. In: CVPR Proceedings (2017). https://arXiv.org/pdf/1611.07004v1.pdf
100. Jaderberg, M., Simonyan, K., Zisserman, A., Kavukcuoglu, K.: Spatial transformer networks (2015). https://arXiv.org/abs/1506.02025
101. Jin, X., Xu, C., Feng, J., Wei, Y., Xiong, J., Yan, S.: Deep learning with S-shaped rectified linear activation units (2015). https://arXiv.org/abs/1512.07030
102. Johnson, J., Krishna, R., Stark, M., Li, L.-J., Shamma, D.A., Bernstein, M., Fei-Fei, L.: Image retrieval using scene graphs. In: IEEE Conference on Computer Vision and Pattern Recognition (CVPR) (2015)
103. Johnson, J., Hariharan, B., van der Maaten, L., Hoffman, J., Fei-Fei, L., Zitnick, C.L., Girshick, R.: Inferring and executing programs for visual reasoning (2017). https://arXiv.org/pdf/1705. 03633.pdf
104. Karpathy, A., Joulin, A., Fei-Fei, L.: Deep fragment embeddings for bidirectional image sentence mapping (2014). https://arXiv.org/pdf/1406.5679.pdf. Accessed on 04 Aug 2017
105. Karpathy, A., Fei-Fei, L.: Deep visual-semantic alignments for generating image descriptions. IEEE Trans. Pattern Anal. Mach. Intell. **39**(4), 664–676 (2017)

106. Kim, H.-R., Kim, Y.-S., Kim, S.J., Lee, I.K.: Building emotional machines: recognizing image emotions through deep neural networks (2017). https://arXiv.org/pdf/1705.07543.pdf
107. Kingma, D.P. Welling, M.: Auto-Encoding Variational Bayes. (2014) CoRR: https://arXiv.org/abs/1312.6114
108. Kingma, D.P., Ba, J.L.: Adam: a method for stochastic optimization. In: International Conference on Learning Representations, vol. 113 (2015)
109. Kiros, R., Salakhutdinov, R., Zemel, R.: Multimodal neural language models. In: Proceedings of the 31st International Conference on Machine Learning (ICML) (2014)
110. Kiros, R., Zhu, Y. Salakhutdinov, R. Zemel, R. S., Torralba, A. Urtasun, R. Fidler, S.: Skip-thought vectors. In: NIPS Proceedings (2015)
111. Krishna, R., Zhu, Y., Groth, O., Johnson, J., Hata, K., Kravitz, J., Chen, S., Kalantidis, Y., Li, L.J., Shamma, D.A., Bernstein, M.S., Fei-Fei, L.: Visual genome: connecting language and vision using crowdsourced dense image annotations. Int. J. Comput. Vis. **123**(1), 32–73 (2017)
112. Krizhevsky, A., Hinton, G.E.: Using very deep autoencoders for content-based image retrieval. In: European Symposium on Artificial Neural Networks ESANN-2011, Bruges, Belgium (2011)
113. Krizhevsky, A., Sutskever, I., Hinton, G.E.: ImageNet classification with deep convolutional neural networks. NIPS **2012**, 1097–1105 (2012)
114. Lai, H., Pan,Y., Liu,Y., Yan, S.: Simultaneous feature learning and hash coding with deep neural networks (2015). https://arXiv.org/pdf/1706.06064.pdf. Accessed on 18 Aug 2017
115. Lample, G., Chaplot, D.S.: Playing FPS games with deep reinforcement learning (2016). https://arXiv.org/abs/1609.05521
116. Larochelle, H., Hinton G.E.: Learning to combine foveal glimpses with a third-order Boltzmann machine. In: NIPS 2010 (2010)
117. LeCun, Y., Bengio, Y.: Convolutional Networks for Images, Speech, and Time Series. The Handbook of Brain Theory and Neural Networks, vol. 3361(10) (1995)
118. LeCun, Y., Bottou, L., Bengio, Y., Haffner, P.: Gradient-based learning applied to document recognition. In: Proceedings of IEEE **86**(11): 2278–2324 (1998)
119. Lienhart, R., Slaney, M.: pLSA on large scale image databases. In: IEEE International Conference on Acoustics, Speech and Signal Processing (ICASSP), vol. 4, pp. 1217–1220 (2007)
120. Lin, T.-Y., Maire, M., Belongie, S., Hays, J., Perona, P., Ramanan, D., Zitnick, C.L., Dollar, P.: Microsoft COCO: common objects in context. In: Computer Vision ECCV Proceedings 2014, pp. 740–755. Springer, Cham (2014)
121. Lippi, M.: Reasoning with deep learning: an open challenge. In: CEUR Workshop Proceedings (2016). http://ceur-ws.org/Vol-1802/paper5.pdf
122. Liu, N., Han, J., Zhang, D., Wen, S., Liu, T.: Predicting eye fixations using convolutional neural networks. In: Proceedings IEEE Conference on Computer Vision and Pattern Recognition (CVPR), pp. 362–370 (2015)
123. Liu, H., Li, B., Lv, X., Huang, Y.: Image retrieval using fused deep convolutional features. Procedia Comput. Sci. **107**, 749–754 (2017)
124. Liu, N., Wang, K., Jin, X., Gao, B., Dellandrea, E., Chen, L.: Visual affective classification by combining visual and text features. PLoS ONE **12**(8): e0183018 (2017). https://doi.org/10.1371/journal.pone.0183018
125. Liu, Y., Zhang, D., Lu, G., Ma, W.-Y.: A survey of content-based image retrieval with high-level semantics. Pattern Recogn. **40**, 262–282 (2007)
126. Long, J., Shelhamer, E., Darrell, T.: Fully convolutional networks for semantic segmentation (2016). https://arXiv.org/abs/1605.06211
127. Lowe, D.G.: Distinctive image features from scale-invariant keypoints. Int. J. Comput. Vis. **60**(2), 91–110 (2004)
128. Ma, L., Lu, Z., Shang, L., Li, H.: Multimodal convolutional neural networks for matching image and sentence. In: Proceedings of 2015 IEEE International Conference on Computer Vision (ICCV) (2015)

129. Ma, L., Lu, Z., Li, H.: Learning to answer questions from image using convolutional neural network. In: Proceedings of the Thirtieth AAAI Conference on Artificial Intelligence (AAAI-16) (2016)
130. Maas, A.L., Hannun, A.Y., Ng, A.Y.: Rectifier nonlinearities improve neural network acoustic models. In: Proceedings of ICML, vol. 30 (1) (2013)
131. Maas, A.L. Hannun, A.Y. Ng, A.Y.: Rectifier nonlinearities improve neural network acoustic models. In: ICML Workshop on Deep Learning for Audio, Speech and Language Processing (2014)
132. Maillot, N., Thonnat, M.: Ontology based complex object recognition. Image Vis. Comput. **26**(1), 102–113 (2008)
133. Malinowski, M., Rohrbach, M., Fritz, M.: Ask your neurons: a deep learning approach to visual question answering (2016). https://arXiv.org/abs/1605.02697
134. Mao, J., Xu, W., Yang, Y., Wang, J., Yuille, A.L.: Explain images with multimodal recurrent neural networks (2014). https://arXiv.org/pdf/1410.1090.pdf. Accessed 06 Aug 2017
135. Mao, J., Wei, X., Yang, Y., Wang, J., Huang, Z., Yuille, A. L.: Learning like a child: fast novel visual concept learning from sentence descriptions of images. In: ICCV Proceedings, pp. 2533–2541 (2015)
136. Mao, J., Xu, J., Jing, Y., Yuille, A.: Training and evaluating multimodal word embeddings with large-scale web annotated images. In: NIPS 2016 Proceedings. http://papers.nips.cc/paper/6590-training-and-evaluating-multimodal-word-embeddings-with-large-scale-web-annotated-images
137. Mansimov, E., Parisotto, E., Ba, J.L., Salakhutdinov, R.: Generating images from captions with attention. arXiv:1511.02793v2 [cs.LG]. Accessed on 29 Feb 2016
138. Menegola, A., Fornaciali, M., Pires, R., Avila, S., Valle, E.: Towards automated melanoma screening: exploring transfer learning schemes (2016). https://arXiv.org/pdf/1609.01228.pdf
139. Mezaris, V. Strintzis, M. G.: Object segmentation and ontologies for MPEG-2 video indexing and retrieval. In: International Conference on Image and Video Retrieval, CIVR 2004. Image and Video Retrieval, pp. 573–581 (2004)
140. Mikolov, T., Corrado, G., Chen, K., Dean, J.: Efficient estimation of word representations in vector space. In: Proceedings of the International Conference on Learning Representations (ICLR 2013), pp. 1–12 (2013)
141. Mesnil, G., He, X., Deng, L., Bengio, Y.: Investigation of recurrent-neural-network architectures and learning methods for spoken language understanding. In: Interspeech (2013)
142. Mikolajczyk, K., Schmid, C.: A performance evaluation of local descriptors. IEEE Trans. Pattern Analy. Mach. Intell. **27**(10), 1615–1630 (2005)
143. Mikolov, T., Sutskever, I., Chen, K., Corrado, G.S., Dean, J.: Distributed representations of words and phrases and their compositionality. In: NIPS Proceedings (2013)
144. Mnih, V., Kavukcuoglu, K., Silver, D., Graves, A., Antonoglou, I., Wierstra, D., Riedmiller, M.: Playing Atari with deep reinforcement learning (2013). https://arXiv.org/abs/1312.5602
145. Mnih, V., Heess, N., Graves, A., Kavukcuoglu, K.: Recurrent models of visual attention. In: NIPS Proceedings (2014). https://arXiv.org/abs/1406.6247
146. Mohamed, A., Dahl, G.E., Hinton, G.: Acoustic modeling using deep belief networks. IEEE Trans. Audio Speech Lang. Process. **20**(1), 14–22 (2012)
147. Mosbah, M. Boucheham, B.: Matching measures in the context of CBIR: a comparative study in terms of effectiveness and efficiency. In: World Conference on Information Systems and Technologies. World CIST 2017, pp. 245–258 (2017)
148. Mozer, M.C.: A focused backpropagation algorithm for temporal pattern recognition. In: Chauvin, Y., Rumelhart, D. (eds.) Backpropagation: Theory, Architectures, and Applications. Research Gate. Lawrence Erlbaum Associates, Hillsdale, NJ, pp. 137–169 (1995)
149. Netzer, Y., Wang, T., Coates, A., Bissacco, A., Wu, B., Ng, A.Y.: Reading digits in natural images with unsupervised feature learning. In: NIPS Workshop on Deep Learning and Unsupervised Feature Learning (2011). http://ufldl.stanford.edu/housenumbers/nips2011_housenumbers.pdf

150. Ng, A.: Sparse autoencoder. CS294A Lecture Notes, Stanford University, Stanford, USA, Technical Report, p. 72 (2010). https://web.stanford.edu/class/cs294a/sparseAutoencoder.pdf
151. Ngiam, J., Khosla, A., Kim, M., Nam, J., Lee, H., Ng, A.Y.: Multimodal deep learning. In: Proceedings of the 28th International Conference on Machine Learning (ICML) (2011)
152. Nguyen, A., Clune, J., Bengio, Y., Dosovitskiy, A., Yosinski, J.: Plug & Play Generative Networks: Conditional Iterative Generation of Images in Latent Space (2016). https://arXiv. org/pdf/1612.00005.pdf
153. Nguyen, G.-H., Tamine, L., Soulier, L.: Toward a deep neural approach for knowledge-based IR (2016). https://arXiv.org/pdf/1606.07211.pdf
154. Noh, H., Hong, S., Han, B.: Learning deconvolution network for semantic segmentation (2015). https://arXiv.org/abs/1505.04366
155. Noh, H., Seo, P.H., Han, B.: Image question answering using convolutional neural network with dynamic parameter prediction. In: 2016 IEEE Conference on Computer Vision and Pattern Recognition (CVPR), Las Vegas, NV, pp. 30–38 (2016). https://doi.org/10.1109/CVPR. 2016.11
156. Novotny, D., Larlus, D., Vedaldi, A.: Learning the structure of objects from web supervision. In: Computer Vision ECCV 2016 Workshops. Amsterdam, The Netherlands, Part 3. LNCS 9915, pp. 218–233 (2016)
157. Pappas, N., Popescu-Belis, A.: Multilingual hierarchical attention networks for document classification (2017). https://arXiv.org/abs/1707.00896
158. Parikh, A.P., Taeckstroem, O., Das, D., Uszkoreit, J.: Composable attention model for natural language inference. In: EMNLP 2016 (2016)
159. Paulin, M., Douze, M., Harchaoui, Z., Mairal, J., Perronin, F., Schmid, C.: Local convolutional features with unsupervised training for image retrieval. In IEEE International Conference on Computer Vision (ICCV), pp. 91–99 (2015)
160. Peng, B., Lu, Z., Li, H., Wong, K.-F.: Towards neural network-based reasoning (2015). https:// arXiv.org/abs/1508.05508
161. Pennington, J., Socher, R., Manning, C.D.: GloVe: global vectors for word representation (2014). https://nlp.stanford.edu/pubs/glove.pdf
162. Perez-Rey, D., Anguita, A., Crespo, J.: Ontodataclean: ontology-based integration and preprocessing of distributed data. In: Biological and Medical Data Analysis, pp. 262–272. Springer, Heidelberg (2006)
163. Peters, R.J., Iyer, A., Itti, L., Koch, C.: Components of bottom-up gaze allocation in natural images. Vis. Res. **45**, 2397–2416 (2005)
164. Petrucci, G., Ghidini, C., Rospocher, M.: Ontology learning in the deep. In: European Knowledge Acquisition Workshop EKAW 2016: Knowledge Engineering and Knowledge Management, pp. 480–495 (2016)
165. Piras, L., Giacinto, G.: Information fusion in content based image retrieval: a comprehensive overview. J. Inf. Fusion. **37**(C), 50–60 (2017)
166. Porello, D., Cristani, M., Ferrario, R.: Integrating ontologies and computer vision for classification of objects in images. In: Proceedings of the Workshop on Neural-Cognitive Integration in German Conference on Artificial Intelligence, pp. 1–15 (2013)
167. Pyykko, J., Glowacka, D.: Interactive content-based image retrieval with deep neural networks. In: International Workshop on Symbiotic Interaction, pp. 77–88 (2016)
168. Redmon, J., Farhadi, A.: YOLO9000: better, faster, stronger (2016). https://arXiv.org/abs/ 1612.08242
169. Reed, S., Akata, Z., Yan, X., Logeswaran, L., Schiele, B., Lee, H.: Generative adversarial text to image synthesis (2016). arXiv:1605.05396v2 [cs.NE]. Accessed on 5 Jun 2016
170. Reed, S., Akata, Z., Mohan, S., Tenka, S., Schiele, B., Lee, H.: Learning what and where to draw. In: Advances in Neural Information Processing Systems 29, Curran Associates, Inc., pp. 217–225 (2016). http://papers.nips.cc/paper/6111-learning-what-and-where-to-draw.pdf
171. Ren, S., He, K., Girshick, R., Sun, J.: Faster R-CNN: towards real-time object detection with region proposal networks (2016). https://arXiv.org/pdf/1506.01497v3.pdf

172. Rezende, D.J., Mohamed, S., Wierstra D.: Stochastic backpropagation and approximate inference in deep generative models (2014). https://arXiv.orabs/1401.4082
173. Ribeiro, R., Uhl, A., Wimmer, G., Haefner, M.: Exploring deep learning and transfer learning for colonic polyp classification. Comput. Math. Methods Med. (2016)
174. Riloff, E.: Automatically generating extraction patterns from untagged text. Proc. Nat. Conf. Arti. Intell. **2**, 1044–1049 (1996)
175. Rosten, E., Drummond, T.: Machine learning for high-speed corner detection. In: Proceeding of European Conference on Computer Vision (ECCV 2006), pp. 430–443 (2006)
176. Saenko, K., Darrell, T.: Unsupervised learning of visual sense models for polysemous word. In: Proceedings of the 22nd Annual Conference on Neural Information Processing Systems. Vancouver, Canada, pp. 1393–1400 (2008)
177. Sak, H., Senior, A., Beaufays, F.: Long short-term memory recurrent neural network architectures for large scale acoustic modeling (2014). https://arXiv.org/abs/1402.1128
178. Salakhutinov, R.: Learning deep generative models. Ann. Rev. Stat. Appl. **2015**(2), 361–385 (2015)
179. Santoro, A., Raposo, D., Barrett, D.G.T., Malinowski, M., Pascanu, R., Battaglia, P., Lillicrap, T.: A simple neural network module for relational reasoning (2017). https://arXiv.org/pdf/1706.01427.pdf
180. dos Santos, C., Tan, M., Xiang, B., Zhou, B.: Attentive pooling networks (2016). https://arXiv.org/abs/1602.03609
181. Sarikaya, R., Hinton, G.E., Deoras, A.: Application of deep belief networks for natural language understanding. IEEE/ACM Trans. Audio Speech Lang. Process. **22**(4), 778–784 (2014)
182. Schawinski, K., Zhang, C., Zhang, H., Fowler, L., Santhanam, G.K.: Generative adversarial networks recover features in astrophysical images of galaxies beyond the deconvolution limit. Monthly Notices of the Royal Astronomical Society: Letters: slx008. https://arXiv.org/pdf/1702.00403.pdf
183. Schuster, S., Krishna, R., Chang, A., Fei-Fei, L., Manning, C.D.: Generating semantically precise scene graphs from textual descriptions for improved image retrieval. In: Proceedings of the Fourth Workshop on Vision and Language, pp. 70–80 (2015)
184. Shench, C., Song, M., Zhao, Q.: Learning high-level concepts by training a deep network on eye fixations. In: NIPS Deep Learning and Unsup Feat Learn Workshop (2012)
185. Shen, C., Zhao, Q.: Learning to predict eye fixations for semantic contents using multi-layer sparse network. Neurocomputing **138**, 61–68 (2014)
186. Shi, J., Tomasi, C.: Good features to track. In: Proceedings of 1994 IEEE Computer Society Conference on Computer Vision and Pattern Recognition, (CVPR, 1994), pp. 593–600 (1994)
187. Simonyan, K., Zisserman, A.: Very deep convolutional networks for large-scale image recognition. In: ICLR (2015)
188. Singh, M.D., Lee, M.: Temporal hierarchies in multilayer gated recurrent neural networks for language models. In: International Joint Conference on Neural Networks (IJCNN) (2017)
189. Smeulders, A.W.M., Worring, M., Santini, S., Gupta, A., Jain, R.: Content-based image retrieval at the end of the early years. IEEE Trans. Pattern Anal. Mach. Intell. **22**(12), 1349–80 (2000)
190. Snoek, C.G.M., Smeulders, A.W.M.: Visual-concept search solved? IEEE Comput. **43**(6), 76–78 (2010)
191. Srivastava, N., Salakhutdinov, R.: Multimodal learning with deep Boltzmann machines. In: NIPS 2012 (2012)
192. Srivastava, N., Hinton, G., Krizhevsky, A., Sutskever, I., Salakhutdinov, R.: Dropout: a simple way to prevent neural networks from overfitting. J. Mach. Learn. Res. **15**, 1929–1958 (2014)
193. Sun, X., Huang, Z., Yin, H., Shen, H.T.: An integrated model for effective saliency prediction. In: Proceedings of Thirty-First AAAI Conference on Artificial Intelligence (2017)
194. Sundermeyer, M., Schluter, R., Ney, H.: LSTM neural networks for language modeling. In: Proceedings of Interspeech (2012)
195. Sutskever, I., Martens, J.: On the importance of initialization and momentum in deep learning (2013). http://doi.org/10.1109/ICASSP.2013.6639346

196. Szegedy, C., Liu, W., Jia, Y., Sermanet, P., Reed, S., Anguelov, D., Erhan, D., Vanhoucke, V., Rabinovich, A.: Going deeper with convolutions. In: CVPR (2015)
197. Szegedy, C., Vanhoucke, V., Ioffe, S., Shlens, J., Wojna, Z.: Rethinking the inception architecture for computer vision (2016). https://arXiv.org/abs/1512.00567
198. Thompson, A., George, N.: Deep Q-learning for humanoid walking. Project advisors: Professor Michael Gennert (2016). https://web.wpi.edu/Pubs/E-project/Available/E-project-042616-142036/unrestricted/Deep_Q-Learning_for_Humanoid_Walking.pdf
199. Tousch, A.-M., Herbin, S., Audibert, J.-Y.: Semantic hierarchies for image annotation: a survey. Pattern Recogn. **45**(1), 333–345 (2012)
200. Town, Ch.: Ontological inference for image and video analysis. Mach. Vis. Appl. **17**(2), 94–115 (2006)
201. Traina, A., Marques, J., Traina, C.: Fighting the semantic gap on CBIR systems through new relevance feedback techniques. In: Proceedings of the 19th IEEE Symposium on Computer-Based Medical Systems, pp. 881–886 (2006)
202. Valle, E., Cord, M.: Advanced techniques in CBIR local descriptors, visual dictionaries and bags of features. In: Tutorials of the XXII Brazilian Symposium on Computer Graphics and Image Processing (SIBGRAPI TUTORIALS), pp. 72–78 (2009)
203. Venugopalan, S., Xu, H., Donahue, J., Rohrbach, M., Mooney, R., Saenko. K.: Translating videos to natural language using deep recurrent neural networks. arXiv preprint arXiv:1412.4729 (2014)
204. Vig, E., Dorr, M., Cox, D.: Large-scale optimization of hierarchical features for saliency prediction in natural images. In: CVPR (2014)
205. Vincent, P., Larochelle, H., Bengio, Y., Manzagol, P.A.: Extracting and composing robust features with denoising autoencoders. In: Proceedings of the Twenty-fifth International Conference on Machine Learning (ICML08), pp 1096–1103. ACM (2008)
206. Vincent, P., Larochelle, H., Lajoie, I., Bengio, Y., Manzagol, P.-A.: Stacked denoising autoencoders: learning useful representations in a deep network with a local denoising criterion. J. Mach. Learn. Res. **11**, 3371–3408 (2010)
207. Vinyals, O., Kaiser, L., Koo, T., Petrov, S., Sutskever, I., Hinton, G.: Grammar as a foreign language (2015). CoRR: https://arXiv.org/pdf/1412.7449.pdf
208. Vinyals, O., Toshev, A., Bengio, S., Erhan, D.: Show and tell: a neural image caption generator. In: CVPR 2015 (2015)
209. Wan, J., Wang, D., Hoi, S.C.H., Wu, P., Zhu, J., Zhang, Y., Li, J.: Deep learning for content-based image retrieval: a comprehensive study. In: ACM International Conference on Multimedia (MM), pp. 157–166. ACM (2014)
210. Wang, C., Zhang, L., Zhang, H.: Learning to reduce the semantic gap in web image retrieval and annotation. In: SIGIR08, Singapore (2008)
211. Wang, H., Cai, Y., Zhang, Y., Pan, H., Lv, W., Han H.: Deep learning for image retrieval: what works and what doesnt. In: IEEE 15th International Conference on Data Mining Workshops, pp. 1576–1583 (2015)
212. Wang, H.: Semantic Deep Learning, University of Oregon, pp. 1–42 (2015)
213. Wang, H., Dou, D., Lowd, D.: Ontology-based deep restricted boltzmann machine. In: 27th International Conference on Database and Expert Systems Applications, DEXA 2016, Porto, Portugal, September 5–8, 2016, Proceedings, Part I, pp. 431–445. Springer International Publishing (2016)
214. Wang, J., Song, Y., Leung, T., Rosenberg, C., Wang, J., Philbin, J., Chen, B., Wu, Y.: Learning fine-grained image similarity with deep ranking. In: Proceedings of the IEEE Conference on Computer Vision and Pattern Recognition (CVPR), pp. 1386–1393. IEEE (2014)
215. Wang, F., Tax, D.M.J.: Survey on the attention based RNN model and its applications in computer vision (2016). CoRR: https://arXiv.org/pdf/1601.06823.pdf
216. Wang, L., Li, Y., Lazebnik, S.: Learning deep structure-preserving image-text embeddings. In: Proceedings of IEEE Conference on Computer Vision and Pattern Recognition (CVPR), Las Vegas, NV, pp. 5005–5013 (2016). https://doi.org/10.1109/CVPR.2016.541

217. Wang, W., Shen, J.: Deep visual attention prediction (2017). CoRR: https://arXiv.org/pdf/1705.02544
218. Wei, Y., Liang, X., Chen, Y., Shen, X., Cheng, M.-M., Feng, J., Zhao, Y., Yan, S.: STC: a simple to complex framework for weakly-supervised semantic segmentation. IEEE Trans. Pattern Analy. Mach. Intell. (2015)
219. Weston, J., Chopra, S., Bordes, A.: Memory networks (2015). arXiv:1410.3916v11 [cs.AI] https://arXiv.org/pdf/1410.3916.pdf
220. Wimalasuriya, D.C., Dou, D.: Ontology-based information extraction: an introduction and a survey of current approaches. J. Inf. Sci. **36**(3), 306–323 (2010)
221. Wu, Y., Schuster, M., Chen, Z., Le, Q.V., Norouzi, M., Macherey, W., Krikun, M., Cao, Y., Gao, Q., Macherey, K., Klingner, J., Shah, A., et al.: Google's neural machine translation system: bridging the gap between human and machine translation (2016). CoRR: https://arXiv.org/abs/1609.08144
222. Xu, K., Ba, J., Kiros, R., Cho, K., Courville, A., Salakhutdinov, R., Zemel, R., Bengio, Y.: Show, attend and tell: neural image caption generation with visual attention (2016). CoRR: https://arXiv.org/pdf/1502.03044.pdf
223. Yang, Z., Yang, D., Dyer, C., He, X., Smola, A., Hovy, E.: Hierarchical attention networks for document classification. In: HLT-NAACL Proceedings (2016)
224. Yin, W., Kann, K., Yu, M., Schuetze, H.: Comparative study of CNN and RNN for natural language processing (2017). https://arXiv.org/pdf/1702.01923.pdf
225. Yosinski, J., Clune, J., Bengio, Y., Lipson, H.: How transferable are features in deep neural networks? In: Advances in Neural Information Processing Systems 27 (NIPS 14) (2014)
226. You, Q., Luo, J., Jin, H., Yang, J.: Building a large scale dataset for image emotion recognition: the fine print and the benchmark (2016). CoRR: https://arXiv.org/abs/1605.02677
227. Yua, S., Jiaa, S., Xu, Ch.: Convolutional neural networks for hyperspectral image classification. Neurocomputing **219**, 88–98 (2017)
228. You, Q., Jin, H., Wang, Z., Fang, C., Luo, J.: Image captioning with semantic attention (2016). https://pdfs.semanticscholar.org/bf55/591e09b58ea9ce8d66110d6d3000ee804bdd.pdf
229. Zeiler, M.D., Fergus, R.: Visualizing and Understanding Convolutional Networks, ECCV 2014. Part I, LNCS **8689**, 818–833 (2014)
230. Zeiler, M. D.: ADADELTA: an adaptive learning rate method (2012). CoRR: http://arXiv.org/abs/1212.5701
231. Zhang, X., LeCun, Y: Text understanding from scratch. eprint arXiv:1502.01710. (2015). CoRR: https://arXiv.org/pdf/1502.01710.pdf
232. Zhang, S., Choromanska, A., LeCun, Y.: Deep learning with elastic averaging SGD. In: Neural Information Processing Systems Conference (NIPS 2015) Proceedings, 1–24. CoRR: http://arXiv.org/abs/1412.6651
233. Zhang, J., Lin, Z., Brandt, J., Shen, X., Sclarof, S.: Top-down neural attention by excitation backprop. In: Leibe, B., Matas, J., Sebe, N., Welling, M. (eds) Computer Vision ECCV 2016. Lecture Notes in Computer Science, vol. 9908. Springer, Cham (2016)
234. Zhang, H., Xu, T., Li, H., Zhang, S., Huang, X., Wang, X., Metaxas, D.: StackGAN: text to photo-realistic image synthesis with stacked generative adversarial networks (2016). CoRR: https://arXiv.org/pdf/1612.03242v1.pdf
235. Zhou, W., Li, H., Tian, T.: Recent advance in content-based image retrieval: a literature survey (2017). CoRR: https://arXiv.org/pdf/1706.06064.pdf. Accessed on 18 Aug 2017
236. Zhu, J.Y., Wu, J., Xu, Y., Chang, E., Tu, Z.: Unsupervised object class discovery via saliency-guided multiple class learning. IEEE Trans. Pattern Analy. Mach. Intell. **37**(4), 862–75 (2015)
237. Zhu, S., Shi, Z., Sun, C., Shen, S.: Deep neural network based image annotation. Pattern Recogn. Lett. **65**, 103–108 (2015)

About the Editors

Professor Dr. Halina Kwasnicka is a Head of Department of Computational Intelligence at Wroclaw University of Science and Technology, Wroclaw, Poland and the Director of Graduate Schools. She was the Deputy Director for Scientific Researches of Institute of Informatics and the head of Division of Artificial Intelligence (in the Institute of Informatics). Over time, her research interest has evolved from nature-inspired methods, data mining, and knowledge-based systems to methods of generation of hierarchies of groups of objects, in the further goal to use them in clustering text documents and images. Such a hierarchy of images projected on ontology could allow inferring semantic content.

Professor Kwasnicka was and is involved in the realization of scientific national and international projects, and published as author or co-author more than 200 journal and conference papers and books.

Dr. Lakhmi C. Jain, PhD, ME, BE(Hons), Fellow (Engineers Australia) is with the Faculty of Education, Science, Technology & Mathematics at the University of Canberra, Australia and Bournemouth University, UK. Professor Jain founded the KES International for providing a professional community the opportunities for publications, knowledge exchange, cooperation and teaming. Involving around 5,000 researchers drawn from universities and companies world-wide, KES facilitates international cooperation and generate synergy in teaching and research. KES regularly provides networking opportunities for professional community through one of the largest conferences of its kind in the area of KES. www.kesinternational.org.

His interests focus on the artificial intelligence paradigms and their applications in complex systems, security, e-education, e-healthcare, unmanned air vehicles and intelligent agents.

Index

© Springer International Publishing AG, part of Springer Nature 2018   163
H. Kwaśnicka and L. C. Jain (eds.), *Bridging the Semantic Gap in Image
and Video Analysis*, Intelligent Systems Reference Library 145,
https://doi.org/10.1007/978-3-319-73891-8

Printed in the United States
By Bookmasters